Essays in Brewing Science

Essays in Brewing Science

Michael J. Lewis and
Charles W. Bamforth
University of California–Davis
Davis, California, USA

Springer

Michael J. Lewis
Professor Emeritus of Brewing Science
Academic Director of Brewing
  Programs in University Extension
University of California–Davis
Davis, California 95616-8598
e-mail: mjlewis@ucdavis.edu

Charles. W. Bamforth
Chairperson, Department of Food Science
  and Technology
Anheuser-Bush Endowed Professor
University of California–Davis
Davis, California 95616-8598
e-mail: cwbamforth@ucdavis.edu

Library of Congress Control Number: 2006923489

ISBN 10: 0-387-33010-0   e-ISBN 10: 0-387-33011-9
ISBN 13: 978-0387-33010-5   e-ISBN 13: 978-0387-33011-2

Printed on acid-free paper.

© 2006 Springer Science+Business Media, LLC
All rights reserved. This work may not be translated or copied in whole or in part without the written permission of the publisher (Springer Science+Business Media, LLC, 233 Spring Street, New York, NY 10013, USA), except for brief excerpts in connection with reviews or scholarly analysis. Use in connection with any form of information storage and retrieval, electronic adaptation, computer software, or by similar or dissimilar methodology now known or hereafter developed is forbidden.
The use in this publication of trade names, trademarks, service marks, and similar terms, even if they are not identified as such, is not to be taken as an expression of opinion as to whether or not they are subject to proprietary rights.

9 8 7 6 5 4 3 2 1

springer.com

Preface

Most brewing texts use a systematic barley-beer-bottle organization that takes the reader sequentially through the various stages of beer-making. This, of course, is logical and useful and works well. However, brewers do not often think about beer and brewing in this way, e.g. to solve problems, but they think about all the stages in the process that might get affected, e.g. a single beer property such as color. Alternatively, brewers might ponder on the influence of such affective agents as modification or oxygen throughout the process. This is also a typical questioning strategy in the examinations of the Institute of Brewing and Distilling that many professional brewers take. We think of this as a longitudinal organization of the subject matter, i.e. looking down the length of the process for causes and effects, and that is the structural approach to this book. It is important to bear this in mind when reading the book because this organization brings together information and ideas that are not usually seen side-by-side, and material that is usually in a single chapter in most books might be spread over several in this one because that best suits the unifying theme of the chapters. It has been relatively easy to draw together, from across the spectrum of beer-making, material that affects such beer properties as color, foam and haze, for example, for which this organizational structure works quite well; however, in other cases the structure has given us some surprises and, to fulfill the concept of the book, e.g. wort boiling appears under a chapter on water and energy. Therefore, though we think of each chapter as a stand-alone essay on the nominated topic, we have been at some pains to cross-reference the chapters one to another, and have worked particularly hard on the index by which every reference to a particular subject can be traced.

The book is not written for an uninformed reader and indeed the approach we have used is inappropriate for those coming to the topic of brewing for the first time. We presume a good deal of knowledge about brewing because our objective is not to teach the fundamentals. For that, a book such as *Brewing* by Lewis and Young (this publisher) would make a good primer. Although we have incorporated the latest ideas from published research and from conversations with researchers and practical brewers alike, we have not peppered the text with references to the original literature; such references tend to "date" a text quickly, yet an individual research paper usually contains such a small kernel of new information that it only makes sense or has relevance when subsumed into the vast storehouse of knowledge that brewers have accumulated over decades and even centuries. Nevertheless, a most inclusive tome such as *Brewing Science and Practice* by Briggs, Boulton, Brookes and Stevens (Woodhead Publishing Ltd., Cambridge) uses references to the original literature and crucially provides an entrée to that source of information for those who need it.

To write a book that is reasonably short on such a vast subject as malting and brewing demands that we make decisions about what to include or exclude and the level of detail that is appropriate in each chapter. Given that each chapter was originally intended to be called an essay and to be about 1000 words in length, our predilection has been to err on the side of brevity. We are sure we will not meet the requirements of every reader in every case. For example, we have made few comments about health issues related to beer though these have increasingly interested the industry; the seminal text in this field is *Beer: Health and Nutrition* by Bamforth (Blackwell, Oxford)) after which there is very little else to say.

Writing this book has taken a good deal of time and arrives well after the date originally promised to the publisher. We therefore thank Susan Safren who chased us relentlessly to complete the book especially when lots of other projects and activities seemed more attractive or more urgent. For the same reason we heartily thank the extraordinarily patient and talented women, whom we were smart enough to marry, for doing all the things they do that made this book possible.

Contents

I Qualities

1. Proteins 3
2. pH 13
3. Color 20
4. Foam 28
5. Haze 43
6. Microbiology 58
7. Inorganic Ions 69

II Processes

8. Raw Materials 77
9. Modification 93
10. Enzymes 105

11. Yeast 114
12. Oxygen 131
13. Water and Energy 143
14. Sanitation and Quality 161

Index 171

Qualities

11

Proteins

Proteins primarily enter the brewing process from barley by way of malt. The bulk of this protein resides within the cells of the endosperm of barley grains where, by the time of grain maturity, the protein forms a matrix in which the large and small starch granules are embedded. There is also some protein in the endosperm cell walls, primarily in the middle lamella that forms the intersection between adjacent cells. Also, there is an intensely hydrophobic protein particle (hydrophobin) present in barleys contaminated with mold spores (notably, Fusarium) that induces gushing of beer and the factor in malt responsible for premature yeast flocculation is proteinaceous in character.

A classical measure of protein is the nitrogen content of grain, which is measured by the Kjeldahl method. The grain is digested completely by boiling sulfuric acid plus catalysts and the ammonium ion, so formed from all the nitrogen-containing substances in the sample, is quantified. Alternatively, the Dumas method can be used, in which the sample is gasified by incineration in oxygen to form oxides of nitrogen; by reduction, the nitrogen gas formed can be measured. In either case the factor 6.25 is then applied to convert total nitrogen to total protein because, on average, proteins contain 16% nitrogen. Because there are many compounds in barley and malt that contain nitrogen but are not proteins, the "protein" value, calculated in this way, is spuriously high and the factor 5.7 is sometimes preferred.

Nevertheless, protein values based on this admittedly false premise comprise a very useful and practical guide for the evaluation of barley and malt. Proteins can also be measured in aqueous samples (wort and beer) by the use of Coomasie Brilliant Blue, a dye that reacts with proteins; analysts read the color formed in a spectrophotometer. Values are always much smaller than the Kjeldahl values for the same samples because the dye is blind to small nitrogen-containing molecules.

Measurement of N-Containing Materials in Brewing

One of the most challenging materials to measure in beer and its raw materials is protein. This is primarily on account of the heterogeneity of the species that are involved. Whereas components such as β-glucan and starch have a single unit building block (glucose), there are 20 or more different monomeric units that comprise proteins, namely the amino acids. Whereas the measurement of these building blocks is straightforward (they react as a group with ninhydrin to afford a violet color and can even be measured individually after fractionating by column chromatography), when they enter into peptide bond formation to produce ever increasing complexities of peptides, polypeptides and proteins, there is a tremendous diversity of species. Simple algebra will illustrate just how many permutations of dipeptide can be constructed from just two different amino acids (20 different amino acids in all manner of permutations, with amino acid 1 providing its $-CO_2H$ group to the peptide bond in one set but its $-NH_2$ group in the other, e.g., Glycyl-alanine and Alanyl-glycine are different dipeptides.) Going successively to tripeptides right the way through polypeptides complicates the situation exponentially.

One way to obtain a reading for total peptides and polypeptides would be to totally hydrolyze the mixture, say as found in barley, malt, wort or beer, and measure the amino acids released. This is seldom efficient, including from a time perspective. One of the most time-revered approaches has been that of Kjeldahl, wherein the totality of nitrogenous materials is converted to ammonia by digestion, and the ammonia measured colorimetrically. As nitrogen constitutes some 16% of the total weight of a protein, the value for N obtained was traditionally multiplied by 6.25 to arrive at an estimation of protein. This method has now been superseded for safety reasons by the Dumas method in which there is total combustion of protein prior to assessment of nitrogen. This gives even higher apparent levels for protein, because of the more comprehensive digestion, but this only illustrates the vagaries of these methods. When the Dumas method was adopted the shortsighted believed that protein levels in barley were actually higher, when in fact a different yardstick was being applied. One difficulty is that the methods do not solely register protein, but other material that contain N, such as nucleic acids and their digestion products.

There are colorimetric methods for assessing protein, so approaches such as the Folin–Ciocalteau method and the Bradford method (Coomassie Blue staining) have been used for estimating protein in beer. However all such methods are dependent on certain types of amino acids being present, and any variation in these will give different intensities of reaction. Simple assessment of ultraviolet absorbance by wort and beer can be used with regression equations to arrive at estimates for protein, but other ultraviolet-absorbing materials such as hop bitter acids interfere.

In common with many grains, the cells of the barley outer endosperm contain perhaps three times as much protein as the innermost endosperm cells. Protein is mobilized to some extent in normal malting, especially during the germination stage, and feeds the growing embryo. Low molecular weight and water-soluble materials accumulate, however, because breakdown of proteins outstrips the rate of utilization of amino acids by the growing embryo. Protein breakdown products contribute to the cold-water extract (CWE) of malt, the nitrogenous component of which is the primary source of free amino nitrogen (FAN, including amino acids) that appear in wort (see Chapter 9). At the end of malting the rootlets or culms are removed from malt kernels taking with them a significant amount of protein and so, generally, barleys and the malts made from them contain roughly the same amount of protein (as N% × 6.25); however, as noted, malt contains protein breakdown products as well as protein.

Malting barley typically contains about 1.8% to 2% total *nitrogen* (or 11.25% to 12.5% *protein*). However, this is a compromise amount and some brewers might prefer, e.g., barley of 1.5% nitrogen (about 10% protein) and others of 2.2% nitrogen (about 14% protein). Low nitrogen content in barley predicts more extract yield from the malt, potentially more chill-stable beer and beers that fine better with isinglass finings, e.g., cask-conditioned ales; in contrast, very low nitrogen barleys tend to be insufficiently vigorous in germination and to contain too little enzyme for many modern mashing regimes. High nitrogen content predicts higher enzymic power (e.g., diastatic power (DP)) in the malt, a factor of central importance in modern rapid processing; excessive nitrogen is undesirable, however, because it lowers extract yield, impedes extract recovery, may contribute unwanted, e.g., haze-forming, proteins to wort and beer and cause the "steely" quality of malt endosperm. The low nitrogen content of approved malting barley varieties is, therefore, a prime indicator of suitability for malting.

Barley proteins are divisible into two general kinds: those that are soluble in salt solution and those that are insoluble in salt solution. The soluble proteins are called albumins and globulins and form about 30% or so of the total protein. They contain, among other things, the enzymes of barley, including β-amylase, for example. The salt-insoluble protein fraction contains the storage proteins, hordein and glutelin. These proteins contain a relatively large amount of proline and glutamine but differ in solubility from each other; hordein is soluble in hot ethanol and glutelin in strongly alkaline solutions of, e.g., sodium hydroxide. These proteins are rendered more easily soluble by the reducing agents mercaptoethanol or dithiothreitol, suggesting they are cross-linked through –S–S– bridges. As the nitrogen content of barley increases the extra nitrogen is preferentially laid down as the less soluble storage protein, hordein. From hordein come polypeptide fragments that are primary contributors to beer haze and to beer foam stability. Barley proteins are characteristic of the variety. Separation of proteins, or protein fractions as enzymes, by, e.g., gel electrophoresis can be used to identify barleys when there is doubt of provenance.

Proteins are fundamentally simple structures being merely unbranched chains of L-α-amino acids linked through peptide bonds; this is the primary structure. Only the amino group, N, on the α-carbon, C, of each amino acid and the adjacent carboxyl atom, C, are involved in the peptide bond and so the backbone chain reads (N–C–C=N–C–C=N–C–C=N–C–C), in which "NCC" represents each amino acid and "=" represents the peptide bond (it has some double bond characters). Each protein molecule has a free-amino end and a free-carboxyl end. Proteins are constructed in the ribosomes of cells and have a specific sequence of amino acids in the primary structure determined by information from the DNA of the cell. As they exit the ribosome, protein chains twist and fold in complex ways to establish their most stable state, called the secondary and tertiary structures, such as an α-helix or a β-pleated sheet; no part of the molecule has random structure. The imino acid proline is located at sharp bends in protein chains. This three-dimensional structure is held together by intramolecular bonds such as hydrogen bonds (between N–H and O–H), ionic bonds, hydrophobic bonds and —S – S— bridges where the local geography of the twisted and folded structure permit such bonds to form. The biological function of proteins, e.g., as enzymes (see Chapter 10) or in structures, e.g., barley cell walls, depends on this native three-dimensional structure; loss of structure, called denaturation, means loss of function. In brewing denaturation happens primarily as a result of heat; brewing processes might be managed to delay or minimize denaturation, e.g., as in kilning of green malt or in mashing to conserve enzymes, or alternatively, to promote denaturation as in kettle boiling (trub formation).

Proteins in beer have two main roles: some proteins help to stabilize beer foam (see Chapter 4) and other proteins react to form hazes in finished beer (see Chapter 5). To be present in beer, proteins must survive the intense reactions of the malting and brewing processes. During germination, proteins of barley are subject to the action of a battery of enzymes that, given time, could reduce them entirely to amino acids. These enzymes arise in the aleurone layer and are released into the endosperm. There are many endopeptidase enzymes that are, for the most part, sulfhydryl enzymes. They differ in the specific site of attack within the chain of amino acids. Endopeptidases attack within the protein molecules and break down 40% to 50% of the protein present to polypeptides. These, still relatively large, molecules are further acted upon by exoenzymes, the carboxypeptidase and (probably to a lesser extent because of a high pH requirement) by aminopeptidase enzymes of malt during barley germination. They attack at the extremities of polypeptide chains to form smaller peptides and amino acids. Amino acids so formed feed the embryo and support its growth. However, the rate of protein breakdown in the endosperm exceeds the rate of synthesis of proteins in the embryo; as a result, significant amounts of low-molecular-weight nitrogenous compounds, measured as FAN (free amino nitrogen), accumulate in malt. FAN is the primary source of amino acids for yeast nutrition in wort. There are no standard methods for the measurement of proteolytic enzymes in malt.

The battery of proteolytic and peptidolytic enzymes in germinating malt is complex and complete enough to ultimately breakdown all barley proteins to their amino acid components; germination is cut short by kilning, however. During kilning, to remove water and fix the properties of malt, profligate enzyme destruction (protein denaturation) results. As a result, the enzymic make-up of kilned malt is quite different from that of green malt in terms of the *kinds* and *amounts* of enzymes present. The proteolytic enzymes are particularly heat sensitive and are unlikely to survive much kilning. Proteins that are not already reduced in molecular size during malting are, therefore, not likely to be much further degraded in mashing. Carboxypeptidases, however, are more heat stable and are present in malt; they are likely to produce some amino acids from peptides during mashing. However, mashing in this modern age is of short duration and done at relatively high temperature and so the opportunity for extensive amino acid production in mashing is less these days than perhaps it once was; aminopeptidases have an alkaline pH optimum and are unlikely to be efficacious in mashing. The enzyme–substrate system is much more dilute in mashing than in malting, which further militates against extensive proteolysis and hence amino acids formation at this stage. Inhibitors of proteolytic/peptidolytic enzymes are released in mashing; finally, the rapid denaturation and precipitation of proteins out of solution during mashing also impedes significant protein and polypeptide breakdown. Brewers generally, therefore, should not consider mashing a useful point of *formation* of amino acids but rather a point primarily of amino acids *extraction*.

Protein-Degrading Enzymes from Malted Barley

Enzyme	Mode of action	Products
Endopeptidase (proteinase)	Hydrolyzes peptide bonds within proteins, polypeptides and larger peptides	Peptides
Carboxypeptidase	Hydrolyzes peptides and polypeptides starting at the carboxyl terminus, releasing one amino acid at a time	Amino acids
Aminopeptidase	Hydrolyzes peptides and polypeptides starting at the amino terminus, releasing one amino acid at a time	Amino acids

Amino acids have a crucial role to play in yeast nutrition. Little or no fermentation takes place in the absence of yeast growth (see Chapter 11), and little or no yeast growth takes place in the absence of assimilable nitrogenous materials. Though yeast is quite able to utilize ammonium salts for growth as a sole source of nitrogen, in brewers' wort yeast is well supplied with organic nitrogen compounds in the form of amino acids derived directly and primarily from malt protein and specifically malt FAN. The content of FAN as ordinarily measured should not be less than 150 mg/l and is usually in the range of 150 to 250 mg/l in wort, depending, of course, on original gravity and amount of adjunct used. Particular amino acids present in wort have the potential to directly affect beer flavor through the formation of specific higher alcohols by deamination and decarboxylation (see Chapter 11).

During the low-temperature stand of mashing, malt proteins (and polyphenols) dissolve, initially, and this continues through the early stages of the mash temperature program. However, as the mash temperature approaches the conversion temperature (about 60°C — 70°C) the proteins react with the polyphenols present and precipitate substantially. These proteins are separated with the spent grains and account for the fact that spent grains contain some 30% (crude) protein (dry basis). This makes brewers grains a desirable material for compounding into animal feed. Protein precipitation might also affect the efficiency of lauter run-off because the precipitated protein–polyphenol–polysaccharide complex is deposited *among* the spent grain particles and could obstruct wort flow in lautering. When significant amounts of air (oxygen) is entrained in the mash, as in some forms of infusion mashing, a more complex reaction can occur.

Most of the protein deposited in the brewhouse is precipitated during mashing. Although the formation of hot break or trub during kettle boiling is visually obvious and quite dramatic, the actual amount of protein substance in the trub flocs is less than it appears to be, and boiling, depending on the duration of the boil, removes perhaps no more than 10% of the total protein removed in the brewhouse. In fact, during mashing there is a great reduction in both the *amount* and *kinds* of proteins present in wort. In boiling the *amount* of protein is somewhat reduced, but there in no further elimination of protein species. This does not mean that protein precipitation by boiling and formation of hot break or hot trub is unimportant; it *is* important, but boiling is not the primary place of protein/polyphenol deposition in quantitative terms. pH affects the extent of protein precipitation in boiling. Normal wort pH is probably close to the isoelectric point of the precipitated proteins and hence to the point of minimum protein solubility because higher or lower pH than normal reduces break formation. Kettle finings (carrageenan) do not increase the amount of protein precipitated but promote formation of flocs or aggregates that are more easily removed, e.g., by sedimentation or in the whirlpool. Hot trub contains not only protein and polyphenol, but also significant carbohydrate material, plus some lipid and minerals.

Upon cooling of wort, protein is further precipitated as cold break (or cold trub). Wort aeration or oxygenation may have a role to play in this through polyphenol oxidation. Some cold break always appears regardless of the duration of boiling. It has the same general composition as hot break (hot trub) and is usually removed from wort by sedimentation (sometimes after yeast addition) in a brink or prefermenter. In the latter case the yeasted wort might be "dropped" to the main fermentation vessel after 12 to 24 hours in a so-called "dropping" system; this leaves behind a deposit made up substantially of break material and ineffectual yeast (dead cells and yeast that flocculates prematurely).

The protein content of beer does not normally change during fermentation, though prolonged exposure to low fermentation temperature might be expected to precipitate some minor amounts of protein/polyphenol complexes that separate with yeast. Protein can be lost in the foam head above the fermentation to the detriment of beer foam (see Chapter 4). In contrast, yeast death and autolysis have the potential to release proteins to beer. Protease release might cause damage to foam proteins especially in high-gravity brews. It is also possible that lowering of pH associated with yeast action could redissolve proteins of cold break (trub) carried forward to the fermenter.

Doubtless, however, there is continuing reaction among protein, polyphenol, metal ions, lipid and polysaccharide during fermentation. When the beer is ultimately cooled in normal finishing processes, e.g., to minus 2°C or so, this material aggregates and settles out (given enough time) or can be filtered out. Isinglass finings are occasionally used to flocculate yeast, but also can serve to aggregate and settle protein. In addition, protein can be removed by adding silica gel to beer often in-line en route to the final filter. This appears to preferentially remove haze proteins leaving foam-positive proteins intact. Other techniques for protein removal, such as partial hydrolysis with a protease (papain), precipitation with tannic acid or bentonite adsorbtion, are some available tools, but these techniques are rapidly losing favor.

Isinglass

One of the great beer genres (viz., the English cask ale) emerged on the backbone of a "natural" clarification process rooted in a protein preparation called isinglass. It is obtained from the dried swim bladders (some call them "maws") of certain warm-water fish, amongst them the sturgeon, catfish, jewfish, threadfish and croaker.

The fish are primarily caught for food use and the functional property of the maw represents added value. Actually, the bladder is even more likely to end up in a Hong Kong soup than as finings for the beer and wine industries.

The bladders are removed, washed and dried. At the smallest scale in a fishing village the maws are sun-dried, but modern fish processing plants use commercial dryers.

Dried maws are ground up, washed and sterilized before being "cut" for a period of around 6 weeks by weak acids such as sulfurous acid and tartaric acid to disrupt the structure of the collagen molecules so as to solubilize the protein and to generate the optimum molecular weight and balance and orientation of positively and negatively charged sites that are responsible for its functionality. Positive charges on the protein attract negatively charged yeast to produce complexes that settle out readily. The negative charges on the isinglass attract positively charged polypeptides and precipitate those. Isinglass functions best with a rising temperature regime in the beer. Isinglass also has lipid-binding capability and is believed to benefit foam stability for this reason.

Isinglass is a very pure form of collagen, the same protein that is found in skin. The reason why isinglass works rather better than collagen from animal hides can be traced to subtleties in its structure. These capabilities were probably first noticed when people stored beverages in bladders as receptacles.

Overview of Protein Impacts

Grist	1.	High N fertilizer increases protein content of grain
	2.	Barley varieties differ in the extent to which they accumulate protein: six-row varieties contain more protein
	3.	Most adjuncts do not contain much protein so act as protein diluents (rice-based adjuncts, corn-based adjuncts, flours, syrups and sugars)
	4.	Approximately 50% of the hordein is solubilized during germination of barley through action of endoproteinases
	5.	Proteinolysis during germination contributes to release of β-amylase in active form from blocking by Protein Z (40,000 mol. wt. protein)
	6.	Carboxypeptidases amply present in raw barley and are further increased during germination. Proteinases synthesized de novo during germination
	7.	Amino acids released through action of endoproteinases and carboxypeptidases react with sugars during kilning to generate color (melanoidins) and flavor (Maillard reaction)
	8.	Mole for mole, the proteins from wheat have superior foaming performance as compared to those from (malted) barley
Sweet wort production	1.	Proteinolysis limited in sweet wort production because native inhibitors kept separate from enzymes in intact grain are released during milling
	2.	Significant precipitation of protein in mashing
	3.	Substantial release of FAN by carboxypeptidase during mashing. pH too low for aminopeptidases
	4.	Peptides released by proteinases and to a limited extent some amino acids provide significant buffering potential
	5.	Oxidation of gel proteins causes cross-linking, this contributing to teig formation and a slowing of wort separation. Hydrogen peroxide produced in this reaction is a substrate for peroxidases in their oxidative polymerization of polyphenols. Latter precipitates more protein as turbidity and affords increased color.
Boiling	1.	Denaturation of proteins increases their hydrophobicity and therefore foam stability and insolubility (hot break formation)
	2.	Removal of proteins through addition of Irish Moss
Hops/hop products		
Yeast and fermentation	1.	Cold break formation on chilling and through oxidation with introduced oxygen
	2.	Substantial loss of protein with yeast head—suppressed by antifoams
	3.	Proteinases secreted by stressed yeast digest foaming proteins
Conditioning	1.	Precipitation of proteins by chilling
	2.	Addition of isinglass (and auxiliary silicate or alginate finings) to promote settling
Filtration and stabilization	1.	Removal of insoluble proteins by filter aids (or membranes in cross-flow filtration)
	2.	Removal of soluble, haze-forming protein by silica hydrogels/xerogels, tannic acid or papain. Papain damages foam by hydrolyzing foaming proteins
Packaging		Pasteurization destroys yeast proteinases, which survive "sterile filtration"
Final product		Surviving proteinases digest foaming proteins

Proteinaceous or polypeptide materials that survive into finished beer are responsible for haze formation (see Chapter 5) and foam stability (see Chapter 4) and, by reaction with other materials in the beer, could make some contribution to the mouthfeel of beer. In addition, amino acids and peptides are potential nutrients for spoilage micro-organisms and their buffering power might promote relatively high pH in beers making such beers (e.g., all-malt products) more difficult to stabilize in terms of haze and microbes and to have a more satiating flavor character.

|2|

pH

pH is the negative logarithm of the hydrogen ion concentration (more exactly the activity), or algebraically $pH = -\log_{10}[H^+]$ or $pH = \log_{10} 1/[H^+]$. This notation was invented by Sorensen to make very small numbers fit into a more comprehensible range; thus a hydrogen ion concentration of 0.0000007 molar or 10^{-7}M is pH 7.0. In doing so however he introduced an important peculiarity: the pH scale is exponential. Therefore, a solution at pH 6.0 contains *ten-times* less H^+ than at pH 7.0 and a wort at pH 5.2 contains nearly four times as much H^+ as a wort at pH 5.8.

Pure distilled water (concentration 55.5 molar) dissociates slightly but equally into H^+ and OH^- ions; the ion product of water, K_w, $[H^+][OH^-]$, can be measured and is 1.0×10^{-14}. The $[H^+]$ in pure water is therefore 10^{-7}M and by Sorensen's definition pure water has a pH of 7.0 (and a pOH of 7.0) and any solution at pH 7.0 is neutral.

An acid is a proton (H^+) donor and a base is a proton acceptor; a weak acid such as acetic acid (HA, a conjugate acid) dissociates to form a proton plus its conjugate base, acetate, thus:

$$HA(\text{acetic acid}) \rightleftharpoons H^+(\text{proton}) + A^-(\text{conjugate base} = \text{acetate}).$$

This is known as a conjugate acid-base pair. It is an axiom of biochemistry that, at any given temperature, the equilibrium concentrations of the

Measurement of pH

pH is measured instrumentally using a pH meter, which comprises two electrodes, a "measuring" electrode and a "reference" electrode (see Figure 2.1). When the measuring electrode is placed in a solution, sodium ions in the glass of the electrode exchange with hydrogen ions in the solution. The concentration of hydrogen ions on the inside of the electrode (provided by 0.1M hydrochloric acid) remains constant. A potential difference is established across the membrane, which will depend on the difference in the concentration of hydrogen ions on both sides of the electrode.

It is important to control temperature when pH is measured and also to know what temperature a pH value is taken at. Because of increased dissociation of molecules and release of protons as the temperature is increased, pH falls. Thus, for example, the pH measured at 65°C is some 0.35 units lower than when measured at 20°C.

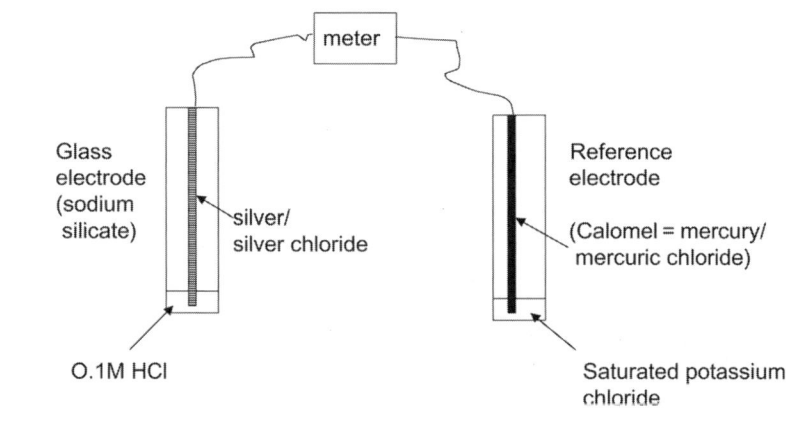

Figure 2.1. pH electrode

conjugate acid–base pair depends only on the pH of the solution. Put the other way around, the pH of a solution (e.g. wort) consisting of conjugate acid-base pairs depends on the ratio of their equilibrium concentrations. This equilibrium can be characterized through the equilibrium constant, K, and the pK_a; the pK_a is defined as $-\log_{10}K$, or more usefully stated, the pK is the pH at which a weak acid is half-dissociated and half-associated. The Henderson–Hasselbalch equation makes use of these ideas and shows that

$$pH = pK + \log_{10}[A-]/[HA].$$

From this the pH can be calculated if the pK and molar concentration of acid [HA] and base [A⁻] are known, or, knowing the concentration of acid and base and the pH, the pK can be calculated. Thus, if a solution be 0.15 molar acetic acid and 0.30 molar sodium acetate then, knowing that the pK_a of acetic acid is 4.8, the pH of the solution will be

$$pH = 4.8 + \log_{10} 0.3/0.15 = 4.8 + \log_{10} 2 = 4.8 + 0.3 = 5.1.$$

The pH of wort or beer is established in a similar way to this example by the relative concentration of undissociated weak acids in equilibrium with their conjugate bases. Note that the most common molecules in wort, sugars and carbohydrates, are uncharged and hence make no contribution to wort pH. Among the many weak organic acids in wort are amino acids; these contain at least two and sometimes three ionizable groups. Thus, if an amino acid such as alanine with pK_a of 2.3 (α-COO⁻ group) and pK_b of 9.9 (α-NH$_3^+$ group) were fully dissociated the pH would be $(2.3 + 9.9)/2 = 6.1$. Most α-amino acids have pKs similar to alanine for the two ionizable group at the α-C atom and contribute strongly to the pH of wort. The common amino acids, glutamic acid and aspartic acid have an additional ionizable –COOH group in the side chain with a pK_a of 4.3 (glutamate) or 3.9 (aspartate); these side chains can ionize even in peptide and polypeptide structures and so contribute additional H⁺ (acidifying) and hence affect wort pH; these groups would be over 90% ionized at wort pH. Similarly, amino acids such as lysine, arginine and others have N-containing structures in the side chain that have pK_bs of 10.8 and 12.5 respectively. At wort pH these structures would be only partially ionized and so serve to soak up or remove some H+ from wort, again contributing to overall wort pH. Thus, interaction of these ionizable groups in organic acids including amino acids and peptides helps to establish wort pH and so e.g. the environment in which enzyme action takes place during mashing.

Another major contributor to wort pH is sodium or potassium phosphate derived also from malt. Phosphoric acid is a weak acid the dissociation of which gives rise to the following equilibrium:

$$H_3PO_4 \rightleftharpoons H^+ + H_2PO_4^- \rightleftharpoons 2H^+ + HPO_4^{2-} \rightleftharpoons 3H^+ + PO_4^{3-} \text{(as } K^+ \text{or } Na^+ \text{salts)}.$$

In practice there is very little H_3PO_4 or K_3PO_4 present in wort; the acid salt KH_2PO_4 dominates in equilibrium with some K_2HO_4. For this equilibrium the pK_a is 7.21. (The three pKs are respectively 2.12, 7.21 and 12.3.)

The weak acid H_2CO_3, carbonic acid, present in brewing water as bicarbonate ions (HCO_3^-), is a major contributor to wort pH; brewers judge the quality of brewing water partly by the amount of bicarbonate ions present (see Chapter 13). The alkalizing (pH-raising and H^+-eliminating) effect of bicarbonate is well known and can be expressed in several ways, e.g.:

$$H_2CO_3 \text{(carbonic acid)} \rightleftharpoons H^+ + HCO_3^- \text{(bicarbonate)} \underset{\rightleftharpoons}{\text{Heat}} CO_2$$

$+ H_2O$ (the pKs are 6.35 and 10.33).

The heat term is included because in relatively dilute and cool solutions an equilibrium is established with minimum alkalizing effect; when hot (as in a mash or in the kettle boil) the CO_2 is driven off and the maximum elimination of H^+ occurs.

$NaHCO_3$ <heat> CO_2 + NaOH (alkalizing reactions) occurs in water that contain an excess of bicarbonate compared to divalent ions (especially Ca^{2+}) and accounts for the unsuitability of such water for brewing.

Calcium ions are present in most brewing waters and are commonly added to water, if needed, in the mash and/or to the kettle usually as the sulfate ($CaSO_4$ or gypsum) or chloride ($CaCl_2$) salt (see Chapter 13). Here the pH-active principle is the Ca^{2+} ion. Ca^{2+} has no acidifying effect of its own, but reacts with phosphate ions, shown in the equilibrium above, particularly with PO_4^{3-}, to form insoluble $Ca_3(PO_4)_2$. This pulls the equilibrium shown to the right with the release of H^+ ions and so engenders a strong acidifying action. As with bicarbonate, brewers evaluate water for its suitable content of calcium or they add calcium salts to achieve some satisfactory level of the ion.

Thus, the pH of wort is the effect of multiple interactions involving the dissociation and association of weak acids and bases (including carboxylic acids, amino acids and phosphoric acid) from malt and reactive ions (mainly bicarbonate and Ca^{2+}) from water. Note that the pH (H^+ content) of water itself has a miniscule effect on wort pH; the pH of water is useful to brewers only as an indicator of contamination with acid or base.

A feature of weak acids or bases is their ability to "buffer," that is to react with H^+ or OH^- to minimize change of pH. They do this most effectively at pH values within about 1.0 pH unit of the pK. In a buffered solution a small addition of H^+ or OH^- has much less effect on the pH of the system than in an un-buffered solution (e.g. plain water); therefore buffering reactions tend to stabilize the pH and it is quite difficult to change the pH of well-buffered systems. Wort and beer are well-buffered systems. Although the pH of a solution is determined by the *equilibrium* of acid–base pairs, the buffering *capacity* depends on the *concentration* of the buffering agents present. Thus, for example, high-gravity all-malt worts are better buffers than low-gravity high-adjunct worts.

The buffering power of any system is greatest close to the pK. Because they have pKs close to the pH of wort or beer, carboxylic acids such as acetic acid (also lactic, malic, pyruvic citric and succinic acids and many others) and the side chain COO^- of glutamate and aspartate and NH^+ of histidine, are important as buffers. Aspartate and glutamate buffer as free amino acids and when linked into peptides and polypeptides because the ionizing group is in the side chain. On the other hand, the α-COO^- and α-NH_3^+ groups of α-amino acids are unlikely to be such good buffers at wort pH because their pKs are generally about 2.0 and 10.0 respectively and, of course, they are substantially bound up in the peptide bonds of polypeptides (see Chapter 1). Phosphate ions (pKs shown above) are also not particularly good buffers in wort and beer.

In a mash the pH ranges from about 5.2 to 5.8, but is normally within one unit of pH 5.4. This pH is measured on cool samples but, because temperature influences pH, the pH of the mash itself might be some 0.3 to 0.4 units lower than that measured on cool worts (depending on the temperatures of the mash and of the measured sample, of course). Cellar wort, therefore, can appear to have a higher pH than brewhouse wort owing merely to this temperature effect. The mash pH is significantly affected by the choice of brewing water (see above) and the choice of malts and adjuncts: generally very hard waters, i.e. Ca^{2+}-containing, yield the lowest pH in the range quoted; darker malts yield a lower pH in the mash than do paler malts, and adjunct mashes tend to have a somewhat higher mash pH than all-malt mashes.

Mash pH is most commonly controlled by the nature of the brewing water or through modification of its composition, by adding, e.g. gypsum (Ca/$MgSO_4$) or an acid such as food-grade phosphoric acid or lactic acid; H_2SO_4 can be used too. An important pH-reducing strategy in some regions where direct acidification cannot be used is a pre-fermentation of a part of the mash with a thermophillic strain of *Lactobacillus* to produce some lactic acid. Enzymes work most rapidly and are longer lived when close to their optimum pH (see Chapter10) and that is the objective of control of mash pH. Nevertheless mash pH was, in the early days of brewing at least, a fortuitous result arrived at, as described above, from the nature of the raw materials used; doubtless this contributed to the historic regional characters of beers. Mash pH is a compromise in which only some enzymes are at their optimum while most others are more or less impeded. Thus, β-amylase has a slightly lower pH optimum than α-amylase and, within a narrow range, lower pH (say pH 5.3 rather than pH 5.5) tends to yield a somewhat more fermentable wort. At such mash pH most other malt enzymes, especially those attacking proteins and polypeptides, are disfavored so that the soluble nitrogen content of wort is somewhat less than could be achieved.

Overview of pH impacts

Grist	1. Higher N of grist, higher is the buffering potential
2. Sugars, syrups and other low N adjuncts lower buffering potential
3. pH reduced throughout malting: increased modification and kilning renders lower pH
4. Use of lactic acid bacteria to lower pH during malting and prevent growth of undesirable organisms
5. Burning sulfur on kiln lowers surface pH—reduces nitrosamine formation
Sweet wort production	1. As mashing temperature is increased, pH decreases
2. pH is 0.1–0.15 lower in worts derived from decoction mashing c.f. infusion mashing
3. Hard water leads to reduced pH through reaction of calcium with phosphate
4. High residual alkalinity (bicarbonates) leads to higher pH
5. pH in mashing may be lowered by adding acid (phosphoric, lactic) or lactic acid bacteria
6. Phytase in malt attacks phytic acid to release phosphate, which in turn reacts with calcium to presage a pH drop.
7. Lowering the pH of a mash by acidification increases phytase action and the released phosphate raises the buffering capacity.
8. When the level of calcium is low there is a sizeable increase in wort pH during run-off, especially as gravity decreases
Boiling	1. The pH of wort drops about 0.3 unit during boiling. Lower gravity worts have a higher pH before boiling, but a substantially bigger pH drop on boiling. Whereas the differences in pH of mashes over the range of gravities 7.5–20° Plato is relatively constant, the final wort pHs is progressively lower as the gravity increases
Hops/hop products	
Yeast and fermentation	1. Addition of acid to yeast in acid washing
2. Organic acids produced during fermentation offer buffering potential
3. Consumption of amino acids and peptides by yeast removes materials that buffer at a higher pH
4. Secretion of protons causes pH drop
5. Any factor that stimulates fermentation causes increased pH drop

	6.	pH rises slightly at the end of fermentation
	7.	High gravity brewing leads to higher pH
Conditioning	1.	Any autolysis of yeast releases buffering material that can raise pH
Filtration and stabilization		
Packaging		
Final product		

Mash pH also affects extraction of color from raw materials and this might well account for different water choices made by brewers of pale beers and dark ones. Generally pale beers are made with water dominated by Ca^{2+} and hence the mash and wort pH tends to be in the low part of the range quoted above. Brewers of dark beers wish to thoroughly extract the color from their expensive special malts and so prefer water with a significant content of bicarbonate; mash and wort pH tend to be high in the range.

|3|
Color

When beer is presented in flint bottles or dispensed in a drinking glass its color conveys important messages to the consumer. Consumers often conflate dark color with stronger flavor impact, higher alcohol content and greater heaviness. Light color conveys the opposite impression. Indeed, if a lager is "colored-up" to look like an ale, consumers are likely to ascribe flavor descriptors normally associated with ales. While interpretations of color by consumers are by no means correct in all cases, beer color and consistent beer color are important quality criteria.

Color arises in beer primarily from the selection of raw materials that comprise the grist, i.e., malts and adjuncts. This color can increase during the kettle boil, the proportion of color from boiling being relatively greater in pale beers than in dark ones. Color can decrease somewhat during fermentation, this effect being greater with dark beers than with pale ones. Color can finally be adjusted to an exact specification by the addition of, e.g., caramels or colored malt extracts, usually postfermentation. Beer color darkens as beer ages.

Color arises in raw materials primarily as a result of the Maillard reaction (named for the French chemist who first described it), also called, descriptively, nonenzymic or nonoxidative browning. In food products, this is a heat-driven reaction between sugars and amino acids to yield highly colored and flavored compounds. Bread crust and the color of toasted bread are perhaps the most common examples of this reaction, and demonstrate clearly that more intense heat yields more intense color in the amber-brown range. These colored products are called melanoidin pigments; they are water soluble and have quite high molecular weight arising from condensations and polymerizations; their chemical structure is unknown though they contain pyrazine and imidazole rings and have high UV absorbance (that provides a method of monitoring early stages of the Maillard reaction) as well as absorbance in the visible range (color). In parallel with pigment formation, flavor compounds are formed, and, although it is possible to separate the color and flavor products of the Maillard reaction on the basis of molecular size, in practice higher malt color also implies higher flavor impact and (as a result of more intense heating) low or zero enzyme content.

Maillard Reaction

Sometimes known as "nonenzymatic browning" these are actually a series of chemical reactions that lead to a brown color when food is heated. The basic reactants are reducing sugars and compounds that contain a free amino group, e.g., amino acids, proteins and amines. There are numerous reaction intermediates and products that include not only color, but also flavorsome compounds and antioxidants. The antioxidants are mostly produced at higher pH values and when the ratio of amino acid to sugar is high. Some of the Maillard reaction products may actually *promote* oxidative reactions. The products of the reaction include Strecker aldehydes, pyrazines, pyrolles and furfurals. Among the flavors contributed are roasted coffee and nuts, bread and cereals. The pyrolles in particular can contribute bitterness. Other Maillard-type reactions occur between amino compounds and substances other than sugars that have a free carbonyl group. These include ascorbic acid and molecules produced during the oxidation of lipids.

The early products in the Maillard reaction are colorless, but when they get progressively larger they become colored. Some of these colored compounds have low molecular weights, but others are much larger, including complexes produced by the heat-induced reactions of the smaller compounds and proteins.

The exact events in any Maillard-based process depend on the proportion of the various precursors, the temperature, pH, water activity and time available. Metals, oxygen and inhibitors such as sulfite also impact. The flavor developed differs depending on the time and intensity of heating; for instance, high temperature for a short time gives a different result to low temperature for a long time. Pentose sugars react faster than do hexoses, which in turn react more rapidly than disaccharides such as maltose and lactose. In terms of amino compounds, lysine and glycine are much more reactive than is cysteine, for instance. The flavor also depends on the amino acid; for example, cysteine affords meaty character, methionine gives potato flavor, while proline gives bready flavor.

As water is produced in the Maillard reaction, the reaction occurs less readily when the water activity is high. And so, it is much more significant in later stages of kilning and curing, and rather less important at the start of killing and during wort boiling.

Malt color is not achieved simply by the application of intense heat to malt that, at a lower temperature, would be pale in color. Maltsters deliberately manipulate the germination (see Chapter 9), and especially the kilning process, to increase the concentration of low-molecular-weight sugars and amino acids in the malt. In kilning this is done by a warm but prolonged early-drying stage so that breakdown of high-molecular-weight substrates continues while the embryo action is slowed or halted by heat and by drying out. This build up of reactants (sugars and amino acids) before application of intense heat promotes the Maillard reaction and hence malt color. If the prolonged early-drying stage is done with little drying, i.e., there is "stewing" on the kiln or in a drum roaster, then this breakdown of endosperm continues such that a mini-mash is conducted inside each kernel. When the malt is then dried the endosperm crystallizes to yield crystal malt and the intensity of heating determines the color of the endosperm. Colored (or caramel) and crystal malts are different in their color and flavor impact, the latter having particularly attractive reddish hues. Barley intended for colored malt manufacture is not necessarily of top grade in every way because manufacture of colored malts result in relatively high malting loss, and high nitrogen might better serve color formation. Gibberellin treatment of germinating barley favors color formation. This can be countered if necessary by application of potassium bromate.

Grist color is substantially extracted during mashing because the melanoidins are water soluble. Color is diluted, relative to original gravity, by use of solid adjuncts, e.g., rice or corn. Mash pH affects extraction, and so those reactions that influence mash pH (water choice and composition, see Chapter 2 and 13, for example) also influence wort color. Generally, higher pH than normal favors more extraction of color from raw materials and also promotes continuing Maillard events. This can be dramatic at extremes of high-mash pH with, e.g., high color pick-up during boiling. Low mash and wort pH helps conserve low wort color. The color of intensely colored malts or roasted materials such as roasted barley might be incompletely extracted in a normal mash, especially as they cannot be milled extremely finely (to promote extraction) lest they impede run-off; this could lead to variability of wort color. It is possible, therefore, to extract such grist materials in a separate process choosing the best temperature and pH (both high) for extraction and creation of color. This extract can be blended to the product, even as late as the finishing cellars, to create the final brown or black beer.

The kettle boil is also a source of wort and beer color, and again the Maillard reaction is responsible because heat drives the reaction between sugars ands amino acids. The effect is, however, rather small in modern short-time boils. Nevertheless, wort pH is important and so pH-changing conditions (e.g., additions such as gypsum, $CaSO_4$ or if bicarbonate ions survive into the kettle) will likely affect these reactions. Addition of sulfur dioxide, often in the form of KMS (potassium meta-bisulfite), will stall the early stages of melanoidin formation by binding to aldehyde groups, and so acts as a bleach; however, SO_2 does not bleach color already formed, and has little or no effect on the Strecker degradation (see below).

Color pick-up in boiling might also increase if significant oxygen entrains as the wort enters the kettle, or if dense first wort is trapped against heating surfaces (e.g., below lower gravity late worts) where it can then scorch and caramelize. Because the Maillard reaction is nonoxidative, the coloring effect of oxygen is probably on reactions involving the polyphenols present to form phlobaphenes; this reaction may be nonenzymic or enzyme catalyzed.

Polyphenol Oxidation

Also known as enzymatic browning, this arises by the oxidation of polyphenols to o-quinones by enzymes such as polyphenol oxidase and peroxidase. Familiar examples are the browning of sliced apple and the making of black tea from green tea.

Whereas heating boosts nonenzymatic browning, the converse applies to enzymatic browning, as the heat inactivates the enzymes. Exclusion of oxygen will also prevent the enzymatic reaction but not the nonenzymatic process.

The most significant enzymes involved in this reaction in brewing systems are the peroxidases. Whereas polyphenol oxidase is present in barley, the level of this extremely heat-sensitive enzyme progressively declines during germination and it is entirely destroyed in kilning and at the very onset of mashing. By contrast, the peroxidases, of which there are several, are extremely heat resistant and present at very high levels in malt. They have very high affinity for hydrogen peroxide and will use very low levels to oxidize polyphenols. The hydrogen peroxide is formed by the reaction of oxygen with thiol compounds in the mash, for example, the sulfhydryl groups located on the gel proteins. In turn, through this reaction the gel proteins stick together as part of the teig formation that leads to a slowing of wort separation.

It needs to be recognized that polyphenol oxidation may also occur nonenzymically, through the agency of activated forms of oxygen, but the enzyme-catalyzed route is more significant.

These complexes are reddish brown in color, but this color can be partially reversed during fermentation by the highly reducing conditions pertaining there. Similarly, simple scorching or caramelization of sugars does not involve nitrogen-containing compounds, although ammonium salts are catalytic to caramel formation. Thermolysis of sugars involves extensive structural changes with dehydration, ring formation (cyclization), condensation and polymerization, forming many different ring compounds. As a result, highly flavorful and colored products are formed. Simple flavorful compounds often result, e.g., maltol, isomaltol and hydroxy-methyl furfural (among many others).

The early stages of the Maillard reaction are known. The basic requirements are water, heat, a primary amine (e.g., amino acids, peptides or even proteins) and compounds with carbonyl groups (e.g., reducing sugars). Pentose sugars, such as xylose and arabinose, react in the Maillard reaction more readily than hexoses that, in turn, are more reactive than disaccharides. Aldose sugars, such as glucose (and, more slowly, ketose sugars such as fructose) in their open chain or free-carbonyl form, condense (i.e., with the elimination of water) with amino acids to form secondary amines (e.g. glycosylamines and, in excess sugar, diglycosylamines). The glycosylamines undergoes in Amadori rearrangement to form a 1-amino-2-keto-sugars or ketosamines. Hydrolysis of such compounds yields 3-deoxyosones (e.g. 3-deoxyosulose from difructoseglycine) and methyl-α-dicarbonyl compounds. Both of these types of compounds, by further reaction with amines, e.g. including amino acids, are the cause of melanoidin pigments that have (among other structures) pyrazine and imidazole rings. There is production also of reductones, which are strong reducing agents, and so the reducing power of colored worts is much greater than pale ones.

One such reaction of dicarbonyl compounds and amino acids is the Strecker degradation in which amino acids react with dicarbonyls to form the

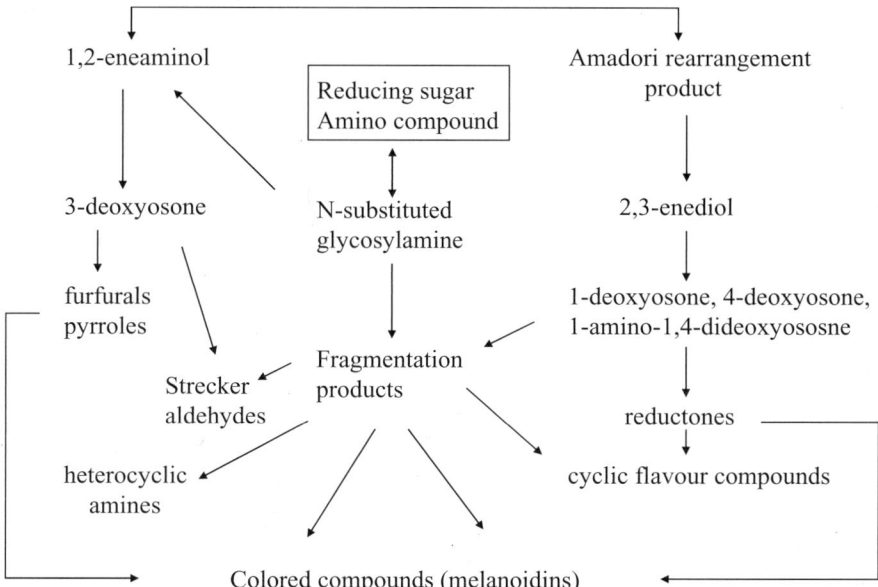

Figure 3.1. The Maillard reaction.

aldehyde (flavorful compounds) that corresponds to each amino acid (plus CO_2) and ketose- or aldose-amine or amino-ketone that corresponds to the sugar reactant(s). These, in turn, can cyclize and condense to form nitrogen-containing pyrazines, pyrroles and substituted pyridines with a great variety of structure and (with sulfur-containing compounds) thiazoles. These have intense flavor. In the Maillard reaction the amino acid and sugar components are ultimately destroyed and are therefore not available for yeast nutrition.

Beer color is commonly measured by two basic and one advanced methods as follows.

(1) *Spectrophotometry*: In this method the absorption of a sample of beer is read at 430 nm. The sample must be perfectly clarified, especially when working with worts, because turbidity causes egregiously high readings. Also, reading absorbance at a single wavelength fails to convey the complexity of color.

(2) *Comparator methods*: In this case, beer color is compared by eye to some standard color source. The Lovibond tintometer, for example, projects on a small screen the color of the beer and the color of standard color glasses. The observer makes the nearest match. Both methods work quite well within a narrow band of products and colors, especially those with which the observer is familiar. However, both methods have poor reproducibility when a broader range of products must be considered. These include problems of the suitability of the color standards and the effects of dilutions of highly colored beers.

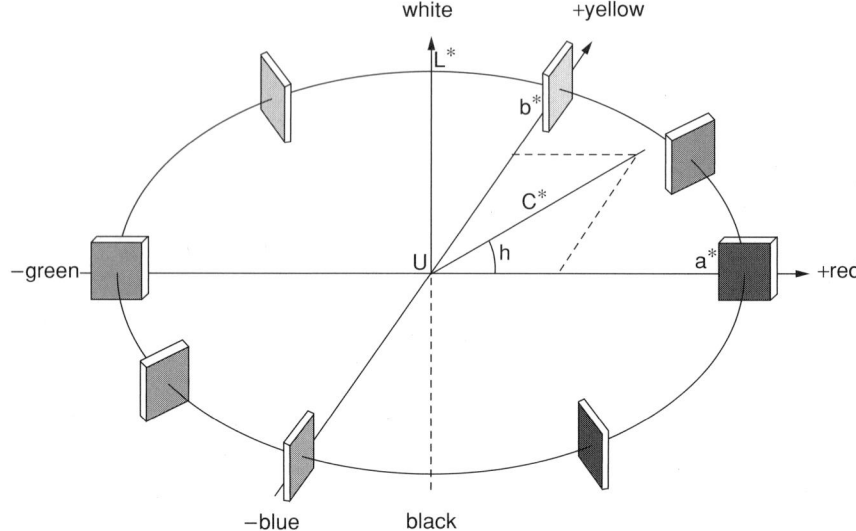

Figure 3.2. Color space (see color insert).

Thus, it frequently happens that two beers with visibly different color quality can yield similar readings when measured in these ways.

(3) *CIE-L*a*b* method*: This method was developed to take into account the fact that color has three qualities (and the method is therefore often called the tristimulus method): hue (color, e.g., red, yellow, etc.), value (brightness or dullness) and chroma or chromaticity (intensity, vividness, or brilliance). The method is based on three values (L^*, a^*, b^*), derived mathematically from the absorbance spectrum of beer in the visible range, the sensitivity of the human eye to different parts of that spectrum and the effect of the light in which the color is viewed. These values can then be used to locate a color, so enumerated, in a three-dimensional space (Figure 3.2) such as a sphere. The equator of such a sphere represents the hue or color; the north and south poles represent color value on a scale of white to black respectively; the length of the radius represents the chroma or chromaticity (vividness of the color). Note that the L^* value is in the range from black (0) to white (100); the a^* value defines color on the red–green dimension and may be a negative or positive value; b^* defines the yellow-to-blue axis, and since beers do not contain blue tones this value is always positive. The color differences between two beers can then be defined as the distance in space between the two points within the sphere that describe their color. By this method a good match is achieved between color as perceived by the human eye and the numerical value of the quality control (QC) method, and the information can be used to accurately adjust final color.

Condiment Brewing

The most efficient processes are the ones that are uninterrupted; traditional malting and brewing operations are anything but that: they comprise a series of batch events and are characterized by variations among those events. For example, a brewery producing, say, 10 different beers may well be using several grist recipes, different mashing regimes, various hop bills, different yeast strains and so on.

A much more efficient (if less aesthetically appealing) approach would be to have a single process stream within the brewery, with one setup only for grist composition, mashing regime, length of boil, etc. Wherever possible there would be continuous operations in place. This is most readily achieved for the fermentation stage—indeed there is a company in New Zealand that operates such a system. There used to be more, but the problem was that such systems operated in the traditional way and did not lend themselves to the production of a range of brands. With the condiment brewing approach such problems no longer arise: the beer coming through the process is one designed to be as bland in appearance and flavor as possible, and it is then adjusted downstream by the addition of materials to produce the range of qualities required.

Thus, isomerized bitter acids can be added to deliver the desired bitterness. Extracted hop oils may be added to introduce late or dry hop character. Color can be introduced by the addition of caramels, but it may be preferred to add extracts of roasted malts. The technology exists to separate according to size the color and flavor in aqueous extracts of such malts: the color molecules are of high molecular weight, the malt flavor molecules are of low molecular weight. Therefore, it is possible to add malty flavors without color and vice versa. Flavors normally provided by yeast, such as esters and sulfur compounds, may be added and the pH can be adjusted by addition of acid or alkali. Salt contents may also be adjusted. Foaming polypeptids can be added.

|4|

Foam

When beer is poured into a glass from a small package or dispensed from a keg, the input of mechanical energy causes gas breakout from supersaturated solution and the bubbles formed rise to the surface. The resulting foam head should be fine bubbled, white and stable. Stable foam is a characteristic of beer that is readily apparent to consumers and a visible harbinger and measure of quality. Beer glasses should always be absolutely clean (commonly called "beer-clean") so that grease on glasses does not cause premature foam collapse.

Many decisions in brewing are a compromise—well illustrated by the compromise between foam stability and haze stability (see Chapters 4 and 5) because proteins (though different ones) drive both phenomena. Thus, in general, factors that improve foam stability aggravate haze. Thus, when discussing foam and haze, it is taken for granted that factors that increase polypeptide content of beer tend to stabilize foam and also increase chill haze potential, as will be seen from the following commentary.

Foam proteins arise in malt; therefore, barley with high protein content might favor sufficient and stable foam. Modification (see Chapter 9) of barley during germination causes extensive protein breakdown, but the amino acids and small peptides that might result do not stabilize foam. Undermodified malt might therefore be preferred because partial hydrolysis of hordeins favors survival of foam-stabilizing polypeptides. By similar argument,

Beer Contains Many Different Polypeptides

It is sometimes forgotten that beer does not contain one or even a very few types of polypeptide. Rather, it contains many different polypeptides that reflect the myriad of changes that occur during malting and brewing: different balances of protein types in the starting grist materials; various extents of hydrolysis by a range of enzymes; a range of degrees of denaturation of polypeptides during wort boiling.

In considering foaming, it is overly simplistic to consider polypeptides in isolation. That a certain polypeptide type should be particularly effective when assessed in isolation in a laboratory does not necessarily mean that it is free to exert its influence on foam in a beer.

Recent research indicates that the net foam quality on beer reflects the levels in beer of polypeptides derived by the hydrolysis of hordeins during germination and the amount of polypeptides that originate in the lipid transfer proteins and protein Z within the albumin components of the grain, as well as the extent to which the latter are denatured. All of these polypeptides are capable of providing stable foams, but the albumin-derived polypeptides are superior. However, it seems that the hordein-derived polypeptides are more adept at entering the foam and, therefore, are able to interfere with the foaming capabilities of the albumin-derived polypeptides. Thus, it is the relative quantities of these two families of polypeptides that are important, more so than the absolute levels of each.

---

unmalted cereal adjuncts, such as raw barley or wheat, are potentially foam positive; but, on the other hand, low-protein adjuncts, such as corn or rice or some syrups, should dilute the protein content of wort and hence support poorer foam. If extensive proteolysis were to occur, e.g., in a long, slow and relatively low-temperature mash, it would militate against better foam performance and for haze stability. Similarly, commercial enzymes are never pure and often contain proteases, possibly foam negative. Although lipids are mostly eliminated from wort and beer by deposition in spent grain, with trub, on yeast and by reaction with lipid-binding protein, foam-negative lipids might be extracted from malt and adjunct if they are excessively milled, mashed very hot with agitation and if the wort is aggressively separated, especially if cloudy worts should result. Malt proteins are precipitated not only in mashing, but also in boiling, and this is, therefore, potentially foam negative. However, boiling makes possible the formation of new foam-stabilizing complexes in wort related to the denaturation of proteins (see Chapter 1), and reactions among groups such as polypeptides, polyphenols, hop acids and inorganic ions and even lipids in the highly reducing conditions of the wort boil. Hops contribute iso-α-acids to wort that are famously foam positive, and also promote lacing or cling (see Chapter 8). However, excessive boiling converts some hop compounds to forms such as humulinic acid that do not support good foam; seeds in hops might contribute lipids. Any foaming or fobbing of beer during processing downstream from the kettle, e.g., overfoaming in fermentation or fobbing during beer transfer, will serve to remove foam-positive materials (particularly polypeptides and hop iso-α-acids) into the fob from which the bulk liquid cannot recover them. Collapsed foam sometimes shows up in beer as skeins or bubble "skins." In a similar way, hydrophobic materials will tend to associate with any surfaces available, such as yeast, filter aids or beer-contact surfaces; very large vessels, therefore, doubtlessly favor foam stability over very small ones. Thus, the interaction among proteins and other potential components of foam, the malting and brewing processes and the formation and stability of foam itself are complex and multidimensional in nature.

Pure liquids do not form foams. Foam is an emulsion of gas in liquid that contains a soluble surfactant. The gas is called the dispersed phase and the thin layer of liquid that separates the bubbles, called lamellae, is the continuous phase. Both elements, i.e., the gas and the liquid, contribute to the stability of the foam formed. Thus, in beer, a dispersed phase that contains pure carbon dioxide is intrinsically less stable than a foam that contains some air or nitrogen because these gases escape the bubble more slowly than CO_2. Similarly, an all-malt beer (as the continuous phase) with somewhat greater viscosity, real extract and significant iso-α-acid and protein content (surfactant), is likely to support a more stable foam than a low-gravity and high-adjunct beer.

Foams have a large interfacial area and are intrinsically unstable because surface tension of the liquid tends toward a minimum value, and this favors foam collapse. Foam collapses by three main mechanisms and two minor ones.

(1) *Drainage* is the downward flow of beer from the foam during foam formation and after the head is established. It results from the effect of gravity. Drainage leaves behind thinner lamellae that rupture more easily contributing to coalescence and disproportionation (see below). As the foam drains, the bubbles lose their spherical shape and, being more compressed together, become polyhedral, and so tend to drain faster. Foam shrinks as a result of drainage, and the relative gas content increases (i.e., the foam becomes drier); as a result, the foam flows less easily and appears to be less creamy. Factors that resist drainage

The Relationship Between Protein Concentration and Foam Stability

As the protein concentration of a beer increases, so does the foam stability, but only up to a point (see Figure 4.1). Eventually a saturation point is arrived at, above which there is no benefit from adding more protein. This is because there is more than sufficient protein to saturate the bubble wall. The nature of the protein is, of course, important, and an excess quantity of protein with a particular ability to enter into the bubble will "squeeze out" the less able protein (which is exactly analogous to competitive inhibition in enzymes (see also the box "Beer Contains Many Different Polypeptides"). Simply speaking, however, the more malt present or the more adjuncts replete with proteins (such as wheat or barley-based adjuncts), the greater the certainty that the beer has ample amount of polypeptide. Beers made with high levels of adjuncts that do not provide protein (e.g., those based on corn, rice or cane sugar), are potentially short of foaming polypeptide and therefore more susceptible to interference by foam-negative agents such as lipids and detergents.

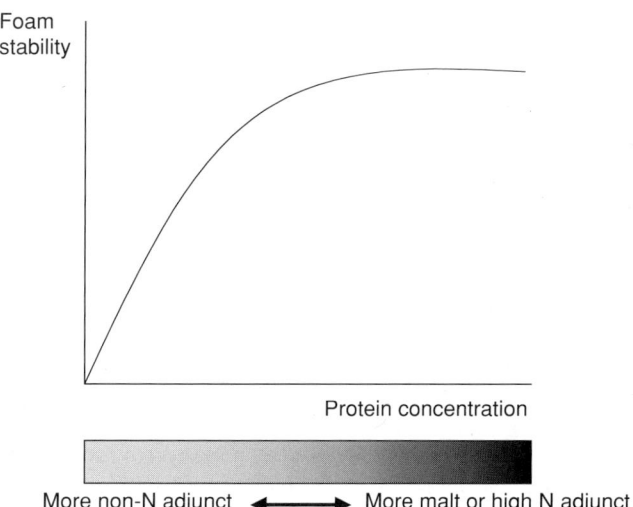

Figure 4.1. The impact of protein concentration on foam stability.

Foam Physics

The key physical events involved in foam formation and retention are as follows:

(1) Bubble formation
(2) Creaming (bubble rise)
(3) Disproportionation
(4) Drainage

They do not occur independently of one another. For example, the extent of drainage of liquid from foam impacts on the tendency for disproportionation to occur. Disproportionation (also known as Ostwald ripening) is the passage of gas from a small bubble to an adjacent larger one, ultimately resulting in the collapse of the former and the growth to unattractive dimensions of the latter.

Foams comprised uniformly of small bubbles are more stable and more appealing. Therefore, the formation of bubbles represents a critical stage for not only the aesthetic appeal, but also longevity of foam.

Bubble Formation

Beers are supersaturated solutions of carbon dioxide, and yet bubbles do not nucleate spontaneously. A nucleation site must be present, which may be a particle, a scratch on the glass or a preformed microbubble. The factors governing the size of bubble that is generated are given by the equation

$$\text{Bubble radius} = [3R_m\gamma/2\rho g]^{1/3}$$

where

R_m = radius of nucleation site (m)
γ = surface tension (mN m^{-1})
ρ = relative density of the beer (kg m^{-3})
g = acceleration due to gravity (9.8 m s^{-2})

Thus, the radius of the nucleation site has a major role to play, but neither the range of surface tensions likely to be encountered in commercial practice (perhaps 42 to 47) nor the range of relative densities, is likely to be of a magnitude sufficient to have sizeable impact.

Creaming

Creaming, often termed "beading," is important not only as an appealing spectacle in beer, but also because it replenishes the foam if it is sustained through the time for which the beer is in the customer's glass.

Ordinarily, the amount of carbon dioxide present in a beer is the overriding parameter governing beading, as is shown by the equation

$$a_n^0 = 3.11C + 0.0962\gamma - 218\rho + 216$$

where

a_n^0 = initial nucleation activity
γ = surface tension
ρ = density
C = carbon dioxide content (vol CO_2/vol beer)

Drainage

As soon as foam is formed, liquid starts to drain from it. The phenomenon is not simple in a medium such as beer; for example, interactions occur between surfactants as the liquid film thins and thus salient parameters such as localized viscosity change with respect to time. However, we can use a simple formula that explains liquid drainage from foams:

$$Q = 2\rho g q \delta / 3\eta$$

where

Q = flow rate (m^3 s^{-1})
η = viscosity of film liquid
ρ = density
q = length of Plateau border (m)
g = acceleration due to gravity
δ = thickness of film (m)

Viscosity is comfortably the most significant factor for drainage, with surface viscosity as opposed to bulk viscosity being most relevant.

Disproportionation

This phenomenon, described above, is governed by the De Vries equation:

$$r_t^2 = r_0^2 - 4RTDS\gamma t / P\theta$$

where

r_t = the bubble radius at time t
r_0 = bubble radius at the start
R = the gas constant (8.3 J K^{-1} mol^{-1})
T = absolute temperature (K)
D = the gas diffusion coefficient (m^2 s^{-1})
S = the solubility of the gas (mol m^{-3} Pa^{-1})
γ = the surface tension
t = time (s)
P = pressure
θ = the film thickness between bubbles

This equation explains the enormous benefit that low levels of nitrogen gas have on foam stability of nitrogen. This gas is much less soluble than carbon dioxide and so less able to dissolve in the liquid interface between bubbles and passing from one bubble to the next. Film thickness is also important, and this will be impacted not only by drainage rates (see earlier), but also by any surface-active materials that enter into the bubble wall and interact to achieve a framework capable of maintaining film integrity (see Box "Foam Model").

therefore affect foam quality (as distinct from foam stability), and so viscosity is a positive attribute of the continuous phase in foams; however, beer viscosity (possibly arising from β-glucans, pentosans and other carbohydrates of real extract, and promoted by low temperature) by itself, is probably a minor contributor to overall beer foam stability.

(2) *Disproportionation* is the diffusion of gas from one bubble to another as a result of gas solubility in the continuous phase and slight differences in internal pressure among bubbles of different size (measured as the radius of bubble surface curvature). Smaller bubbles have higher internal gas pressure than larger ones, and so gas diffuses from small bubbles (that shrink) into larger bubbles (that grow). As a result, the foam coarsens in appearance and the bubbles collapse quickly. Carbon dioxide foams collapse much quicker than when air or nitrogen is in the gas phase because CO_2 has a much higher solubility in the continuous phase than air or N_2. A foam head formed of small bubbles of even size tends to be stable because disproportionation is minimized. The bubbles at the surface of foam are a special case. Here, foam bubbles can leak gas to the atmosphere, not merely to another bubble. If the gas in the bubble is pure CO_2, there is a steep diffusion gradient between the bubble and the atmosphere and, because CO_2 is readily soluble in beer, the gas leaks rapidly causing

foam collapse. However, air from the atmosphere diffuses into the surface bubbles (though only slowly because it is poorly soluble in beer). As a result of these mechanisms, foam shrinks and eventually leaves a residual thin foam. However, if the gas in the bubble dissolves slowly in beer and is amply present in the atmosphere (so that there is a small concentration gradient driving diffusion), the foam will be stable. These factors explain why air (in hand pump dispense) or nitrogenization of beer is effective in stabilizing foam.

(3) *Coalescence* is an increase in bubble size as a result of one bubble collapsing into another; excessive thinning of the lamellae among bubbles causes this, and is hence related to foam drainage. As a result of coalescence, the foam coarsens and collapses as larger but fewer bubbles appear. Contamination of foam from greasy glasses (or greasy lips or facial hair) causes foam collapse by coalescence, especially at the foam surface. As the contaminant spreads, it drags along liquid in the lamellae causing local thinning and, hence, rupture of the bubbles. By such mechanism, a beer might be entirely devoid of even a residual foam in a few minutes after dispensing.

(4) *Beading* is the continuous formation of new bubbles, by nucleation of gas bubbles within the serving glass, and their release into the bulk of the liquid; they reinforce the foam head with fresh bubbles from below. To form a bubble in this way, a significant energy barrier (surface tension) must be overcome; supersaturation of gas in beer and a nucleation site, such as a scratch in the surface of a glass, or a particle in the beer itself, are essential for this to happen.

(5) *Evaporation* of water from the surface of beer foam, much affected by beer temperature, considerably accelerates foam collapse. Lids on traditional beer steins, along with dispense of cold beer, probably limit evaporation and help establish a CO_2 atmosphere above beer and so protect foam.

Beers contain surfactant molecules that permit the formation and stabilization of a large surface area of liquid spread over gas bubbles. The surfactant molecules prevent bubble collapse and stabilize the foam; these are proteins and iso-α-acids primarily, but also some metal ions and viscous materials. Beer also contains foam-destabilizing agents (mostly lipids). The proteins of beer derive from malt and survive in beers in degraded and denatured forms, which are best called polypeptides; these undoubtedly are the most important beer components that stabilize foam. Various forms of polypeptides have been ascribed major roles in foam stabilization, such as glycoproteins, high molecular weight proteins, lipid-transfer protein and hydrophobic polypeptides. It is only this last category that addresses the essential quality that a surfactant (surface tension reducing) material must have: the innate ability to associate with the enormous surface formed during foaming and thus separate into foam. Such proteins must have (or be able to acquire) hydrophobic (water hating) as well as hydrophilic properties. In this way, suitable polypeptides can act at the interface between the gas and the liquid of the foam and stabilize it. Surfactants also tend to remain in the lamellae when drainage takes place and thus tend to increase in concentration. The iso-α-acids, e.g., tetra-hydro-iso-α-acid (which is particularly foam positive), are also hydrophobic molecules that tend to separate into the foam where they serve to stabilize foam and promote cling or foam lacing. This accounts for the fact that foam is always more bitter than the beer from which it is derived. Indubitably, iso-α-acids and proteins act cooperatively in foam stabilization and these two components might well react together, even in the foam itself, to form new effective surfactant molecules. Metal ions, especially iron and copper, also separate into foam and, in the presence of hop substances, might well be part of the surfactant complex that stabilizes beer foam. Beer viscosity (as opposed to localized viscosity in the foam itself) may have some positive effects on foam quality (e.g., creamy appearance), but more likely has a minor effect on foam stability.

Foam-negative factors are lipid in nature, mostly fatty acids. These could be derived from malt and other grist ingredients such as cereal adjuncts though, as noted above, only a very small portion of malt lipids survive brewing and fermentation to form beer. Release of lipids from yeast, particularly autolysing dead cells, could be a major contributor of lipids. Perhaps for this reason, beers fermented at high gravity with a high pitching rate, appear to be less foam stable than those brewed at normal gravity. However, the release of proteolytic enzymes might be an alternative explanation for this observation because yeast tends to produce more proteinases when stressed under conditions of high-gravity brewing, e.g., high osmotic pressure, high-alcohol environment and through aging. This militates against using yeast for too many generations and abjures unnecessarily prolonged exposure of beer to yeast. Because these proteases are destroyed by pasteurization, unpasteurized beers are more susceptible to loss of foam quality in trade than are pasteurized beers. Furthermore, processing of high-gravity worts in the brew house, especially kettle boiling, could precipitate more foam proteins than at normal gravity.

Curiously, beer appears to bind-up and carry harmlessly (at least as far as foam is concerned) small amounts of lipid. Upon addition of a fatty acid to beer, foam stability at first decreases and after some hours, returns to normal. This is likely due to the lipid-binding protein. It is not inconceivable that lipoproteins, such as hydrophobic/hydrophilic surfactants formed as an ordinary part of brewing processes, might have a foam-stabilizing role.

The most important qualities of foam are its appearance (it should be white and fine-bubbled) and its stability or longevity. There are many reported methods of foam measurement, none being entirely satisfactory. Most

Foam Model

A simple model to explain the interactions that occur in the bubble wall is shown in Figure 4.2. Hydrophobic interactions occur between the hydrocarbon side chains of the iso-α-acids and amphipathic polypeptides. Furthermore, divalent metal cations link together adjacent bitter acids through ionic bonds—i.e., the two positive charges effectively neutralize the two negative charges contributed by the iso-α-acid anions. And so, the model explains why the hydrophobic nature of polypeptides is so important for foaming, as is the extent of hydrophobicity of the bitter acids (remembering that reduction of these acids increases hydrophobicity). It also explains why divalent metal cations such as iron, copper and zinc will promote foam stability.

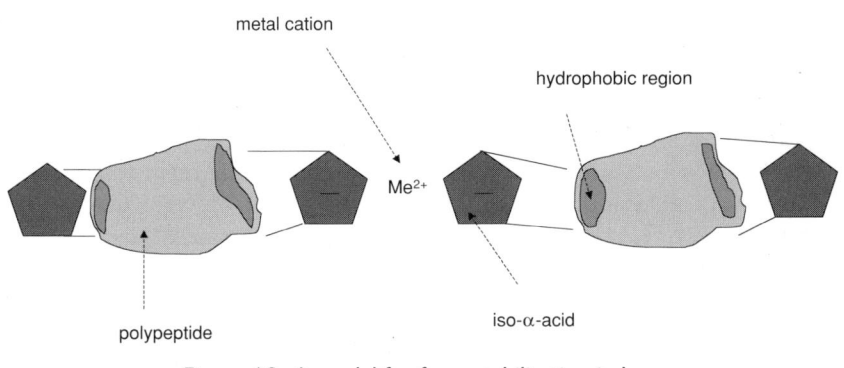

Figure 4.2. A model for foam stabilization in beer.

methods are named after researchers who first described them. The most common methods of foam assessment are based on ideas of Helm, Blom, Ross and Clark, and measure the rate of drainage of beer from beer foam. They differ only in the means by which foam is formed and drainage is assessed. For example, the Rudin method foams up a sample of degassed beer by passing CO_2 through a sintered disc. Time taken for a preset volume to drain is measured. Rudin yields the half-life of the foam as HRV (head retention value) in seconds. The America Society of Brewing Chemists (ASBC) sigma foam method, by contrast, raises the foam by pouring beer into a straight-sided funnel with a bottom tap. The method compares, according to a strict time program, the volume of beer that drains to the volume of beer that remains in the foam and reports a sigma (Σ) value. Some methods trace, by electronic or optical means, the foam–beer interface after pouring or gassing up; these are still drainage methods. One problem with all drainage methods (particularly, perhaps, those that depend on artificial foaming of degassed beers) is that foam drainage is only one of the determinants of beer foam stability, and they tend to overemphasize the role of viscosity in foam stability. Further, drainage is not how foam stability is perceived by consumers; therefore, there is poor correlation between HRV or Σ, for example, and consumers' observations. Nevertheless, most data about beer foams have been collected using such analytical methods and have informed brewers' opinions about foam. Conductivity can be used to estimate foam volume (Blom principle), and electrodes can be used to trace the air–foam interface and thus measure foam collapse directly (Nibem meter); there still remain problems of generating the foam reliably and making reproducible measures that correlate with consumers' opinions.

The most practical method of foam measurement might require that beer be poured normally into a glass and foam collapse defined and rated by direct human observation. This might be perfectly satisfactory in a single QC laboratory and with a known and rather narrow range of beers, but does not meet the needs of researchers or interlaboratory communication, and the data would tend to vary with time and the particular observer. In contrast, the pouring methods of Constant and Yasui measure the depth of the band of beer foam, over time, after pouring. From these data, Yasui calculates an FCT (foam collapse time) value and Constant an NHL (normalized half-life) value for the foam. Neither of these values has any correlation with ASBC Σ-foam values (and likely any other drainage methods) for the same beers; however, both FCT and NHL correlate well with consumers' observations of foam, and also with those components of beer expected to support good foam including, e.g., total protein measured by Coomasie Brilliant Blue. Both these methods depend on simple but reliable beer-pouring machines.

Foam can also be assessed by photographic means, for example to estimate bubble size and distribution or to quantify lacing or cling on beer glasses.

Various agents have been used to stabilize foam. The most effective of these is nitrogen gas combined with a device to initiate gas breakout. For draft dispense this is a suitable orifice plate in the bar tap, or, in packaged goods, a "widget." The downside is the complex technology of packaging such beers, the fact that they must be chilled below their optimal drinking temperature to prevent gushing, the beers take on a mellow and smooth character that is not suitable to all beers and nitrogen suppresses the aroma contribution of hop oils. Propylene glycol alginate (PGA), made by the partial esterification of alginic acid extracted from seaweed, does not directly stabilize foam but protects foam from exogenous lipids, e.g., from beer glasses, by adsorbing them. Divalent metal cations have sometimes been used to stabilize foam; zinc is the preferred ion added to the kettle because it also supports satisfactory yeast performance. Copper, iron, nickel and (notoriously) cobalt also promote foam but cannot be used because they also promote beer oxidation.

Overview of Foam Impacts

Grist	1.	Barley—higher protein means more foam potential
	2.	Proteinolysis during germination of barley causes formation of more soluble hydrophobic (poly)peptide degradation products with high foamability and some foam stability
	3.	Grist also supplies lipid transfer protein (LTP1) and protein Z (40,000 molecular weight protein), which are less foamable than hordein degradation products, but have more foam stability
	4.	Wheat proteins have more foam stability than the equivalent proteins from barley
	5.	Adjuncts based on corn, rice and sugars provide no foaming material and their use, therefore, dilutes foam potential
	6.	Roasted adjuncts may enhance foam through their contribution of melanoidins
Sweet wort production	1.	High gravity brewing leads to poorer foams because extraction of hydrophobic polypeptides is inefficient and there is greater loss of relatively insoluble hydrophobic material at higher concentrations
	2.	Higher temperature mashes favor better foam
	3.	Lower pH mashes (e.g., < 5.1) favor foam
Boiling	1.	Denaturation of proteins exposes their internal hydrophobicity and therefore increases foam stability
	2.	This is balanced with protein lost by precipitation
Hops/hop products	1.	Higher α-acid gives increased foam-stabilizing iso-α-acid
	2.	Reduced iso-α-acids have superior foam stability, but may lead to less appealing foams
Yeast and fermentation	1.	Foaming during fermentation removes foaming polypeptides, and so antifoams enhance finished beer foam provided they are removed on the filter
	2.	Increase losses during high gravity brewing
	3.	Yeast stressed at higher gravities autolyses and releases damaging proteinases
	4.	Therefore, avoid excessive generations of yeast (5 maximum recommended)

Conditioning		Prolonged contact of yeast with beer damages foam through autolysis and proteinolysis
Filtration and stabilization	1.	Overfiltration of beer will lead to excess removal of foaming material
	2.	Papain damages foam polypeptides
	3.	PGA may be added as a foam protectant. Divalent cations such as zinc stabilize foam. Iron has been used, but it promotes oxidation
	4.	"Right first time" preferred, as it avoids correction of gases with attendant foaming and loss of head potential
Packaging	1.	Addition of nitrogen gas increases foam stability
	2.	Use of widget in small pack as a nucleation device
	3.	High carbon dioxide leads to increased foamability
Final product	1.	Mixed gas (CO_2/N_2) dispense of draft beer enhances foam presentation
	2.	High ethanol damages foam
	3.	Lipids and detergents introduced at dispense severely damage foam
	4.	Higher temperature promotes foam formation but decreases foam stability

Troubleshooting Beer Foam Problems: Checks and Balances

This list ranks in categories of likely significance process impacts on foaming excellence.

Top Priority

1. CO_2 and O_2 in specification right first time
2. Original extract/alcohol in specification
3. Bitterness in specification
4. Fermentation under control, enabling residence time in fermenter to be on target and yeast removal within required "window"
5. No more than five "generations" of properly stored and handled yeast
6. Length of boil and percentage of evaporation restricted
7. Minimize all process foaming
8. Outlet checks: age of beer in trade, line and glass washing procedures and dispense setups

Medium Priority

1. Grist: percentage of malt, wheat malt, micronized cereal and crystal malt

2. Highest practical mash-in temperature
3. Mash acidification (pH < 5.1)
4. Avoidance of exogenous enzymes
5. Hop quality (age, alpha)
6. Antifoams
7. Isinglass as lipid binders
8. Clarity of beer ex-cold tank vs. filtration needs: avoid overfiltration
9. Postfermentation bittering (reduced isomerized extracts)
10. PGA—metal ions?
11. Nitrogen gas

Lowest Priority

1. Use of less well-modified malt
2. Avoidance of low temperature rest in mashing
3. Enhancement of clarity of wort entering fermenter

Gushing

Spontaneous generation of foam when a package of beer is opened is called gushing. It can be caused by solid particles in the beer acting as nucleation sites for bubble release, or it can be due to stable microbubbles of gas produced by agitation acting as nucleation sites. The most common cause of gushing is the presence of a small peptide (molecular weight ca. 15,000) called hydrophobin, which is produced by molds such as *Fusarium* that infect grain. This peptide is extremely hydrophobic and forms potent nucleation sites. It is essential that the grain used for making malt is devoid of this type of infection, which is a particular risk for grain grown in wetter climates. Reliable methods do not yet exist for the measurement of hydrophobin. However, if deoxynivalenol (DON), the vomitoxin also produced by *Fusarium*, is present (it can be measured by HPLC [High Performance Liquid Chromatography]), then this is a secondary indicator of gushing potential.

Gushing may also be caused by a range of agents that are capable of nucleating microbubbles and these include the following: crystals of calcium oxalate, slithers of glass in inadequately washed new bottles, rough surfaces on the inside of bottles, heavy metals such nickel, oxidized and dimerized resins in old hops and bittering extracts, filter aid breakthrough, any haze particles and excess carbonation.

If most bottles gush, then this is suggestive of the problem being hydrophobin related. If there is variability between cans or bottles, then this indicates one of the other forms of gushing.

Gushing potential is tested for by applying some form of agitation before allowing settling and then weighing the amount of beer that leaves the opened container.

Grist	Avoidance of barley infected with *Fusarium* and other field fungi
Sweet wort production	Precipitation and removal of oxalate with calcium
Boiling	
Hops/hop products	Avoidance of hop preparations containing oxidation and dimerization products
Yeast and fermentation	
Conditioning	
Filtration and stabilization	1. Avoidance of filter aid breakthrough 2. Avoidance of heavy metals
Packaging	1. Thorough rinsing of new glass to eliminate nucleating particles 2. Good control of carbon dioxide throughout packaging runs
Final product	Avoidance of agitation

|5|

Haze

When beer is poured into a glass from a small package or dispensed from a keg, whether the beer is clear ("brilliant" in brewers' parlance) or cloudy is immediately obvious to the consumer. Though a few beer types, such a wheat beers, are deliberately served hazy (in which case *consistent* haze is an issue), for the most part consumers expect clear beer; they might (rightly) suspect poor quality and reject beer that is not clear. Producing beers that are clear and remain clear in the trade is therefore an essential requirement of almost all brewers.

We have discussed elsewhere (see Chapter 4) that there is a fundamental compromise between beer foam and clarity; proteins (though different ones) drive both phenomena and, in general, factors that improve foam aggravate haze. Thus, it is taken for granted that factors that increase polypeptide content of beer tend to stabilize foam and increase haze potential. Hazes can arise from numerous causes. Although the reaction between polyphenols and proteins is undoubtedly the most common cause of haze in modern brewing and primarily occupies brewers' actions for control (and is the only form of haze addressed here), starch, metal ions, β-glucans, pentosans, hop products, oxalate, foam stabilizers, filter aid (and so on) can also cause so-called nonbiological haze. In addition, the presence of yeast, as a result of bottle conditioning or accidental contamination with wild yeast or bacteria, can cause biological haze (see Chapter 6). In either case, haze is commonly

Pitfalls of Haze Analysis

One of the most difficult tasks facing a brewer when a haze problem arises is identifying the composition of the haze. Usually the dry weight of the haze is relatively small, and so the brewer is obliged to filter or centrifuge enormous quantities of the product in order to render sufficient material for a comprehensive analysis. Such a thorough analysis involves a digestion of the haze in strong acid or alkali, followed by individual measurement of how much protein, carbohydrate (both β- and α-glucan and pentosan) and polyphenol is present, as well as other possible materials such as oxalic acid and heavy metal ions. Usually a haze comprises several, if not all, of these materials. Once a particle has started to be formed in a beer, it is to be expected that any material that is relatively poorly soluble will tend to attach to such a body rather than remain in the aqueous environment of beer.

For practical simplicity in the face of the difficulties of recovering sufficient material, most people resort to simple staining techniques to assess haze material. Thus, agents such as eosin yellow stain protein, Congo red stains β-glucan, iodine stains starch and so on. It must be stressed, however, that the idea is not to measure what is on the surface but to identify the initial cause of the haze, which means what is found at the heart of the haze particle, as it is this which formed the nucleus of the haze problem. In reality this can seldom, if ever, be achieved.

associated with undesirable flavor change; e.g., bacteria can cause beer to sour and protein–polyphenol haze spoils the appearance of beer and is often associated with oxidized flavor.

The primary source of haze-forming materials in brewing is malt. This is the source of specific haze-potentiating proteins and polyphenols. Hops also contribute polyphenols. Brewers therefore select low-protein barleys that are easily modified (see Chapters 8 and 9) for malting, so that the survival of protein into beer is minimized at the outset. It is also possible these days to select barley that has a low content of polyphenol (anthocyanogen-free or ant-free barley) that is highly effective in yielding haze-stable beer. A related strategy for control of such hazes is to use thoroughly well-modified malt, and thus, maltsters' strategies for good modification are a part of the defense against haze. More directly, brewers commonly dilute, by up to 50%,

Bits

Some beers are ostensibly "bright," but if you look closely you can see strands or fibers floating in the liquid. Sometimes they are difficult to discern, but at the other extreme they can appear almost as a snowstorm.

The most famous example of such "bits" was in a major U.S. brewing company in the 1960s. The injudicious use of two separate stabilizing agents led to them interacting with one another to form huge quantities of particles in the beer. The company did not seem to feel it to be a major problem, but the customers did, and the company failed within a very few years.

The example illustrates the usual source of bits—the interactions between agents added to beer. Another example came a few years ago with an alcohol-free beer marketed in the Middle East. It was originally marketed in cans, with no apparent difficulties, but in due course was sold in bottles. It was then that the customers noted with disapproval the copious precipitate that was clearly visible in the bottom of the bottles. When poured into a glass, the precipitate broke up into bits. Clearly in this instance, the customers found the precipitate more objectionable than the bits. The cause of the problem was an interaction at the very high local temperatures, between PGA foam stabilizer and isinglass finings that had been used to clarify the beer. By the simple expedient of eliminating PGA, the problem was solved.

Bits are difficult to quantify using haze meters. One way to assess bits levels semiquantitatively is to recover them by filtering through paper and staining the papers with methylene blue. The filter papers are compared with reference papers generated using artificially generated bits.

---

the malt used in mashing with adjunct materials such as preparations of rice or corn (maize) that are naturally low in protein and polyphenol. Such beers are intrinsically more haze stable than all-malt products.

Brewhouse processes are vital opportunities for the deposition of protein and polyphenol; milling, of course, exposes the husk and endosperm to extraction by brewing water in mashing. Brewers assume excessive milling promotes undesirable extraction of husk polyphenols, but experience with hammer-milled malt suggests that this concern is misplaced. In the early, low-temperature stages of a temperature-programmed mash, protein and polyphenol dissolve from the grain. However, as the mash rises toward

46   Chapter 5

Dead Bacteria

One of the more curious sources of haze in beer is dead bacteria. Haze is perhaps too strong a term to use; it is more a case of a lack of brilliance in the beer. Its occurrence depends on a combination of adverse events. One specific example can be described: malt was being obtained from a less than efficiently cleaned malt house. As a result, substantial quantities of rod-shaped bacteria populated the grain. During kilning, this microbial population was killed, but nonetheless survived on the malt delivered to the brewery. It was washed off the malt in mashing and survived through into the finished beer. The brewery concerned employed perlite and not kieselguhr as a filter aid. The former was less adept at removing the dead bacteria.

Oxygen in the Brewhouse

There has been a major focus in recent years on trying to minimize oxygen pick-up in the brewhouse, in the supposed interest of improving flavor stability of beer. In fact, there seem to be as many reports of oxygen in the brewhouse having no impact on flavor stability as those that say it does. What is less arguable is the fact that oxygen ingress in the brewhouse does impact the colloidal stability of beer. It was Dennis Briggs who first made additions of an "active" form of oxygen, hydrogen peroxide, into mashes to oxidize polyphenols and cause their agglomeration with proteins and removal at the wort-separation stage. As a result, lower levels of haze precursors emerged into the finished wort, and the resultant beers had increased resistance to haze development.

Oxygen entering into a mashing system reacts with the so-called gel proteins. The sulfhydryl side chains in these proteins (provided by cysteine residues) react with the oxygen and, as a result, cross-link (Figure 5.1). The resultant protein agglomerates serve to slow down wort separation as they form a clay-like mass in the grain beds. Hydrogen peroxide is produced

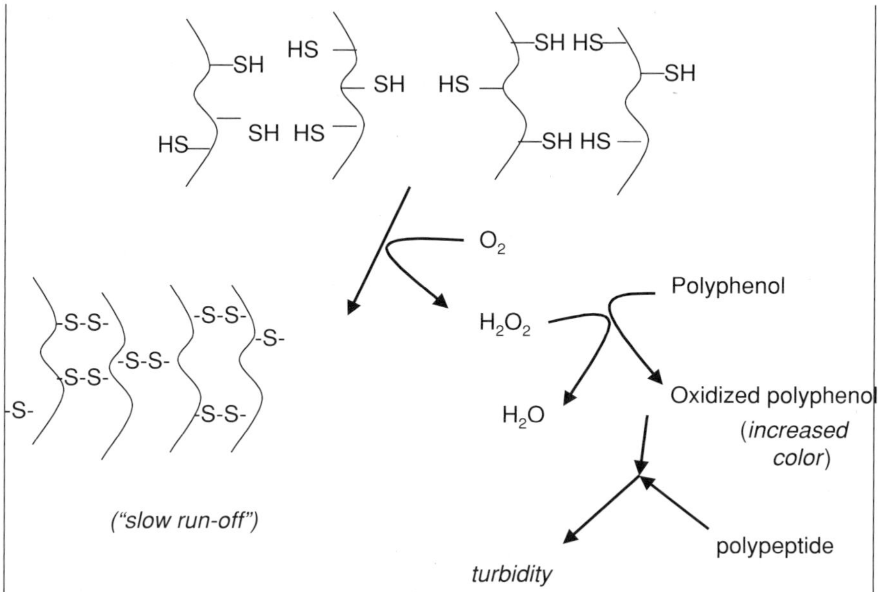

Figure 5.1. Oxidative reactions in mashing.

and this forms a substrate for peroxidase, which catalyzes the oxidation of polyphenols to form red oxidation products (these increase the color of the wort). The oxidized products also cross-link with hordein-derived polypeptides in the wort to form insoluble complexes that can be filtered out. As a result, there is less of these polypeptides and polyphenol left to go forward to the finished beer.

conversion temperature, protein and polyphenol react and proteins substantially (about 80%) precipitate in the mash and so exit the process in the spent grains (which comprises about 30% crude protein, dry weight). Not only the *amount*, but also the *kinds* of proteins present in wort are affected by this precipitation. During wort boiling, more protein–polyphenol complex is precipitated as "hot break." The amount precipitated is a function of a vigorous boil (a "full rolling" boil being essential) and the length of the boil. These days, 45 to 60 minutes at the boil has replaced much longer boils (up to 180 minutes) of former years, and the effect is to decrease the amount of hot break formed. However, the total amount of protein–polyphenol removed in boiling is quantitatively small (though of course qualitatively important) and the *kind* of proteins present is not much changed. Nevertheless, shortened boils, often allied with downstream palliative measures, work satisfactorily for haze prevention. When wort is cooled below about 90°C and then to

Breaks

The insoluble material produced during the boiling of wort is generally called "hot break," and that produced when wort is subsequently cooled is termed "cold break." There are substantial differences between these materials, not least the amount, there being up to five times more hot break than cold break in a well-run operation. Levels of cold break can be between 40 and 350 mg/l. Hot break particles flocculate, but those of cold break do not.

Particles of hot break tend to be rather larger at up to 0.8 cm in diameter, with those of cold break seldom rising above 1 mm in diameter. The proportion of polyphenol and carbohydrate tends to be greater in cold break, while that of protein is higher in hot break. Bitter acids are not found in cold break and lipid is only present in significant quantities in hot break.

Hot break formation appears to be, for the most part, a consequence of protein denaturation, with neither the bitter acids nor the polyphenols playing a positive role in particle formation. Conversely, polyphenol oxidation and resultant cross-linking with proteins seems to impact cold break formation.

---

cellar temperature, the cold break forms. In some brewing systems this is removed by settlement before yeast addition and, in others, after addition of yeast, by a "dropping" system. This is probably a wise strategy because the formation of the hot and cold breaks is affected by wort pH. Wort at its ordinary pH of about 5.2 to 5.4 is at the isoelectric point (i.e.p.) of many proteins (see Chapter 1) present; thus, being electrically neutral, they tend to be least soluble. If the pH is above or below their i.e.p., proteins are more soluble and "breaks" form during boiling to a much smaller extent or even not at all. Because there is a substantial drop in pH during fermentation, there is a good chance that cold-break, if allowed to proceed into the cellar, could compromise beer stability as it redissolves. Nevertheless, the prolonged time and low temperature of fermentation and, especially, finishing processes undoubtedly favor further precipitation of protein–polyphenol complexes.

The last defense of brewers against protein–polyphenol haze (as part of the ordinary maturation and finishing processes) is cold storage. At a

Cold Stabilization: Is Time or Temperature More Important?

Most brewers hold beer postfermentation at a very cold temperature as part of a stabilization regime. Traditionally, a minimum of 3 days storage at −1°C has been advocated, though many brewers store for rather longer than that. It has been shown, however, that the extent of precipitation that can be achieved is actually complete within a comparatively short time at this temperature, indeed in much less than 1 day. Equally, it has been shown that the lower the temperature, the more material is precipitated, so that at −2°C, for example, material that is not precipitated at −1°C is brought out of solution. It is thus the extent of coldness achieved that is more important than the length of time for which the beer is stored.

Although it is largely water, beer does not freeze at 0°C because of the presence of molecules dissolved in it. The higher the original extract/alcohol content of the beer, the more resistant it is to freezing according to the equation

$$\text{Freezing point}(°C) = -(0.42\,A + 0.04\,E + 0.2)$$

in which A is the percentage of alcohol content by weight and E is the original extract of the wort. Therefore, each 1% increase in alcohol content lowers the freezing point by 0.42°C and each increase in extract of 1°P lowers it by 0.04°C. Thus, no beer will freeze at −1°C, and products at higher alcohol concentrations (including high-gravity brews prior to dilution) will withstand even lower temperatures.

---

temperature of −2°C or so (for some days or weeks) the haze material forms and comes out of solution as particulate matter and, at this same low temperature, is removed by settlement or (more efficiently) filtration. If the beer warms up before or during filtration, the effectiveness of this treatment is seriously compromised because the haze complex redissolves; thus beer should see its lowest temperature in the brewery and be well filtered at this temperature.

Protein–polyphenol haze might be characterized as (1) chill haze (that redissolves when the beer is warmed) or (2) permanent haze (that does not redissolve), or can be (3) a mixture of the two (haze that partially redissolves

The Siebert Model

Siebert has described a model for protein–polyphenol interactions (see Figure 5.2). The model assumes that it is primarily proline-containing proteins that interact to form chill haze, with a fixed number of polyphenol-binding sites *in toto*. Furthermore, it is assumed that a polyphenol has two (or more) "ends," which can specifically interact with these binding sites on proteins, thereby allowing a single polyphenol molecule to bridge between protein molecules. If there is an excess of haze-active protein over haze-active polyphenol (and this is usually so for beer), then most polyphenols are involved in bridging two proteins together, with insufficient polyphenol to bridge dimers and form larger particles. If the haze-active polyphenol is in excess of protein (as for instance will occur in ciders), there will be a shortage of free proline sites able to enter into cross-linking of protein molecules. It is only when there are roughly equal quantities of haze-active protein and haze-active polyphenol that the conditions exist for the formation of large networks that will manifest themselves as visible particles. Naturally, the levels of both will need to be sufficiently high to generate a visible haze when they associate.

Such a model has major implications for the stabilization of beer. For instance, "single ended" polyphenols would be expected to block haze formation by competing for proline residues in proteins and preventing cross-linking. Indeed, an excess of either haze-active protein or haze-active

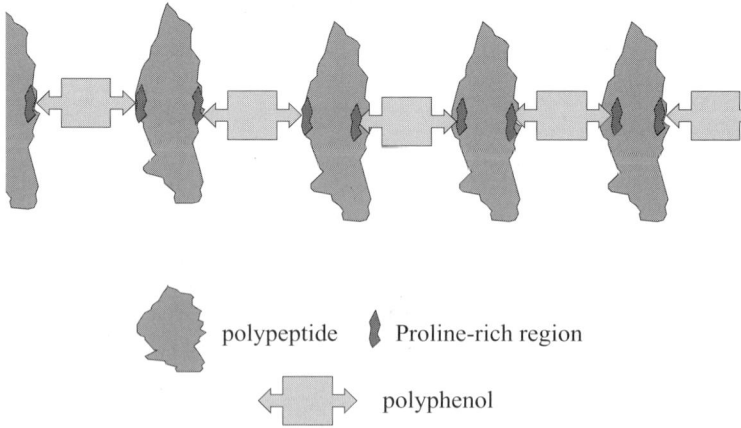

Figure 5.2. The Siebert model for haze formation (see color insert).

> polyphenol would be expected to counter haze development, so one could even rationalize on ensuring that a beer is "over-dosed" on one or the other, but certainly not both.

upon warming the beer). This suggests that proteins and polyphenols initially associate by hydrogen bonds (reversible by heat), which become more permanent valence bonds with time. Haze can be induced in most beers by cycling them between hot (60°C) and cold (0°C) conditions.

Beers are hazy because they contain particles that scatter light. Although beers can produce, say, 20 mg/l of haze material, the amount (weight) of such materials required to render a beer visually unacceptable in the trade is vanishingly small, much less than can be easily isolated and weighed. For this reason light scattering is the practical method of measuring haze. Although most instruments measure light deflected at 90° from the incident light, different instruments are not generally comparable and give different results with different beers; the particle size of the standards used to calibrate instruments, and the particle sizes in the beers being measured, also influence the data. Interlaboratory comparison of haze data (as in all such communications) must therefore be done with care. Beer haze, measured in such instruments, must be compared to haze standards that can be made from a reaction between hydrazine sulfate and hexamethylenetetramine (formazin), or purchased as standards commonly used for the analysis of water. The major analytical groups ASBC and EBC have adopted different scales for haze: 1.0 unit EBC is 69 ASBC units and this value in each case represents "brilliant" (absolutely clear) beer; beers are considered "slightly hazy" and are unlikely to be commercially acceptable above 2 EBC or 140 ASBC.

Invisible Haze

Sometimes beers register high readings on haze meters that rely on measuring light scatter at 90° to incident, even though they appear bright to the human eye. They are said to contain "invisible haze." There was an attempt to call it by the rather more sensible sounding "pseudo haze," but the term invisible haze seems to have stuck. The phenomenon is a problem because it means that brewers must overrule the values that their instrumentation read and make a judgmental decision on whether a beer is or is not acceptable for release to packaging. If a beer was to be refiltered simply on the basis of the haze meter reading, it would not alleviate the situation as the particles responsible for invisible haze are extremely small (< 0.2 µm) and not removed by conventional filtration or centrifuge operations.

There may be several causes for invisible haze. Two origins are particles of unconverted starchy endosperm of barley and cell-surface material sloughed off badly handled yeast.

An alternative way around the invisible haze problem is to make haze readings using a machine that measures light scattered at "forward" angles, e.g., 13°. Such meters are not as sensitive to the very small particles. However, some brewers believe they should know whether these particles are in their beer as they believe that they form the nucleus for the growth of bigger, clearly visible hazes. It is uncertain whether this is the case.

With a reliable and reproducible method of measuring haze, brewers attempt to predict the haze stability of beers because in many markets haze formation is the determining factor of shelf-life (in other markets it is the genesis of oxidized or stale flavor, see Chapter 12). The strategy is to force the beer to form haze by promoting the protein–polyphenol reaction so that months of shelf-life are compressed into a few days. There is generally no problem in promoting hazes in beers; the difficulty is in correlating the accelerated haze with the haze shelf-life observed in the trade. Forcing methods are of two general kinds: first are those that cycle the beers between hot conditions (e.g., 60°C for 48 hours) and then cooled (e.g., 0°C for 24 hours) for measurement by reading the haze after each cycle. Beers that withstand more cycles or form less haze are obviously more haze-stable. Second, beers can be treated with various precipitants, in which the precipitant is titrated into beer. Ethanol (for the alcohol precipitation limit) and ammonium sulfate (for saturated ammonium sulfate precipitation limit (SASPL)) are the most common precipitants used. Both tend to react with the total protein content of beers, and so total protein (e.g., measured with Coomasie Brilliant Blue, not Kjeldahl) also correlates with haze shelf-life.

The approximation used to this point suggests that haze is the result of a reaction between generic protein and generic polyphenol; this does serve to explain the palliative measures that brewers use to control haze formation and to render their products haze-stable. Protein + polyphenol = aggregates = haze, is a simplification that works to demonstrate how haze can be controlled: (1) as described above, in the brewery, brewers permit the haze to be formed and then remove it, and (2) subsequently, brewers can remove either or both reactants to prevent the reaction. These days, silica gel is used to remove proteins from beer and PVPP (polyvinylpolypyrrolidone)

Prediction of Haze Stability of Beer

Colloidal shelf-life may be defined as the length of time before a beer displays a haze value of 2.5 °EBC (175 °ASBC) at 0°C. It is customary for brewers to subject samples of their beer to accelerated aging regimens and see how many cycles of aging a beer can tolerate before registering such a haze value. There are two main types of tests: hot/cold cycling tests and precipitation tests.

Hot/Cold Cycling

There are many different variations of such a test. One involves simple holding of beer at 37°C, in which 1 week at this temperature is said to be the equivalent to 1 month of "normal" storage (18°C). Another uses cycles of storage at 60°C for 2 days/–2°C for 1 day, in which one complete cycle is said to be the equivalent to 6 weeks "normal" storage. Another uses alternating 24 hours cycles of 30°C and 0°C, with one complete hot/cold cycle approximating to one month of "normal" storage.

Precipitation Tests

In the alcohol-chilling test of Chapon, ethanol is added to lower the temperature of a beer to –8°C and chill haze is forced out within a total test time of 40 minutes. This test predicts only chill haze. Alternatively, haze-active proteins can be precipitated by the addition of tannic acid. Gallotannin can be replaced by a solution of saturated ammonium sulfate, and hence it is named the SASPL test. In the first case, the amount of light scatter caused by a standard addition of tannic acid is measured. In the second, the number of milliliters of $(NH_4)_2SO_4$ that need to be added to cause a measurable increase in turbidity is recorded. More salt needed to bring out protein means that less precipitable protein is present in solution. By using polyvinylpyrrolidone monomer (PVP) as precipitant, tannoids can be quantified.

is used to remove polyphenol. Some brewers might choose to use both products sequentially, though combined treatment methods and products are now emerging. These insoluble adsorbents have largely replaced papain, bentonite and tannic acid, for example, which were formerly popular. The new products are preferred because they act rapidly in a beer stream—e.g., on the way to filtration or can be incorporated into filter sheets—and are likely more specific for haze reactants, and so excessive tank bottoms (that were significant, e.g., with tannic acid and bentonite) are eliminated. Some grades of PVPP can be recovered and regenerated with caustic soda on site.

Stabilizers

Various materials can be used for the downstream removal of haze-forming substances.

PVPP adsorbs polyphenols and two types of PVPP are available: (1) single use, which is a micronized white powder with high ratio of surface area to mass, which readily adsorbs polyphenols on its surface and can be incorporated into a filter aid body feed dosing regime; (2) regenerable, which can either be impregnated into sheets or used within devoted horizontal leaf pressure vessels and used after mainstream filtration. These are regenerated by treatment with 1% to 2% caustic.

Silica hydrogels and xerogels are derived from sand. Xerogels are produced from hydrogels by drying before the milling stage that is used to derive the preferred particle sizes. The critical features include the pore size of the particles and the surface area presented by them. A reduction in particle size (giving an increase in surface area) increases the adsorption rate. This is of particular significance for the mode of use of hydrogels. If they are dosed into storage tank, time will allow equilibrium to be established. Conversely, if they are to be dosed in-line as a partial substitute for filter aid, then adsorption rate is an especially important parameter. Silica hydrogels with low permeability (filterability) afford better stability. To achieve beers of prolonged shelf-life, the brewer should employ either low-permeability gels in storage tank, with a commensurate decrease in throughput and increased filter aid usage, or use larger quantities of high-permeability gels. Silica hydrogel and xerogels remove haze-forming protein preferentially to foam-active polypeptide. The silica recognizes and interacts with the same sites on haze-active polypeptides, as do the polyphenols.

It is a moot point, still not wholly answered, whether a competition can exist in beer, in which polyphenols and hydrogels vie for the polypeptides. If this is the state of affairs, then it may be that high levels of polyphenols interfere with stabilization efficiency by silicas, and that a cotreatment of beer with PVPP and silica hydrogel would be best. Newer preparations in the market comprise a combination of the PVP monomer with silica.

Tannic acid is a potent precipitant of haze-active proteins in beer, and it throws a sizeable precipitate when added to a cold conditioning tank. This necessitates either cautious transfer of beer from sediment material in tank or the use of a polisher centrifuge followed by membrane filtration. Alternatively, because the reaction time of new generation gallotannins is so rapid, they can be dosed on-line to the powder filter. Tannic acid is normally added from a 1% to 5% solution made up in deaerated water at room temperature in the brewery. Sedimentation time following dosing on transfer from fermenter to storage tank is at least one day, depending on temperature, yeast count, vessel geometry, etc. Whilst tank bottoms are increased, 25% longer filter runs may be obtained. If the gallotannin is dosed as beer flows to filter, 5 to 10 minutes contact time at 0 to $-1°C$ is necessary and the filter aid must be adjusted to a much coarser grade of kieselguhr or a blend with a high (90% +) proportion of perlite.

Papain, from papaya was the first haze-preventative employed in the brewing industry. It retains some usage, particular by brewers taking a "Belt and Braces" approach for beers that are destined for particularly challenging conditions. However, proteolytic enzymes lessen foam quality by damaging foam polypeptide; however, papain does not require any special equipment to enable its use. Papain is added on transfer to maturation or during maturation itself. It progressively loses its activity during storage and especially during pasteurization. Notwithstanding, it will continue to act for a limited time in a tunnel pasteurizer, and for a short period will effect proteolysis approximately 100-fold more rapidly than during cold conditioning. And in sterile-filtered beers any papain would survive into the package to progressively lower foam stability.

A new enzyme preparation on the market is prolyl endopeptidase, which specifically hydrolyzes peptide bonds involving proline. As it is the haze-forming proteins derived from hordeins that are particularly rich in these residues, the enzyme has more selectivity than papain. It may even be the case that this enzyme helps beer foam. As described in the foam chapter (Chapter 4), hordein-derived proteins seem to be less good foam stabilizers than are the albumin-derived polypeptides, and if the former are removed there is a greater opportunity for the albumin-derived polypeptides to enter into the bubbles.

Newer preparations, now being marketed for the simultaneous removal of polypeptide and polyphenol from beer, are agarose based.

The proteins of haze material primarily arise in the hordein or prolamin (storage) fraction of barley, though some nonprolamins are also present in beer haze. The prolamin fraction increases most in higher nitrogen barley and again explains why low-nitrogen barley is preferred for malting. These alcohol-soluble proteins have a high content of proline, the residue of which seems to be essential for haze formation. They are different from the foam-stabilizing proteins (see Chapter 4). Incidentally, these proteins are also responsible for the immune reaction experienced by celiacs; haze prevention in beer and rendering the beer "gluten free" are therefore compatible practices. Similarly not all polyphenols form haze. Flavanols and proanthocyanidins are the most important haze precursors; dimeric and trimeric forms of catechin, epicatechin and gallocatechin are prime candidates and higher polymers would be even more effective. Though highly polymerized polyphenols are too reactive to survive the brewing process into beer, polymers probably form in beer by oxidation over time, especially in beers with high dissolved oxygen (DO) (see Chapter 12). This probably accounts for the fact that formation of haze and stale flavor has similar kinetics. Studies in model systems suggest that haze is best formed when there is an appropriate balance of protein and polyphenol in solution, because this leads to the largest aggregates that become insoluble most quickly. If either protein-binding sites or polyphenol-binding sites dominate the system, the formation of large aggregates (and so haze) is impeded. Brewers' practical experience that reducing *either* the protein (e.g., with silica gel) or the polyphenol (with PVPP) content of beer confers haze stability, tends to bear out this view. Even with such treatments, however, beers are not haze-*proof*.

Overview of Haze Impacts

Grist	1.	Higher protein grain gives increased level of haze-forming polypeptide
	2.	Low-proanthocyanidin barleys give less haze-forming polyphenol
	3.	Alkaline steeping to remove polyphenols from husk
	4.	Inadequately modified malt gives increased risk of β-glucan/pentosan/starch hazes
	5.	Wheat adjuncts give increased risk of pentosan hazes
	6.	Low-protein adjuncts (rice, corn, syrups, sugars) lower haze potential
Sweet wort production	1.	Inadequate gelatinization and amylolysis leaves unconverted starch
	2.	Low-temperature mash-in enables removal of residual cell wall polysaccharides
	3.	Exogenous glucanases and pentosanases remove residual polysaccharide
	4.	Precipitation of proteins during conversion stage
	5.	Sufficient calcium to precipitate oxalic acid

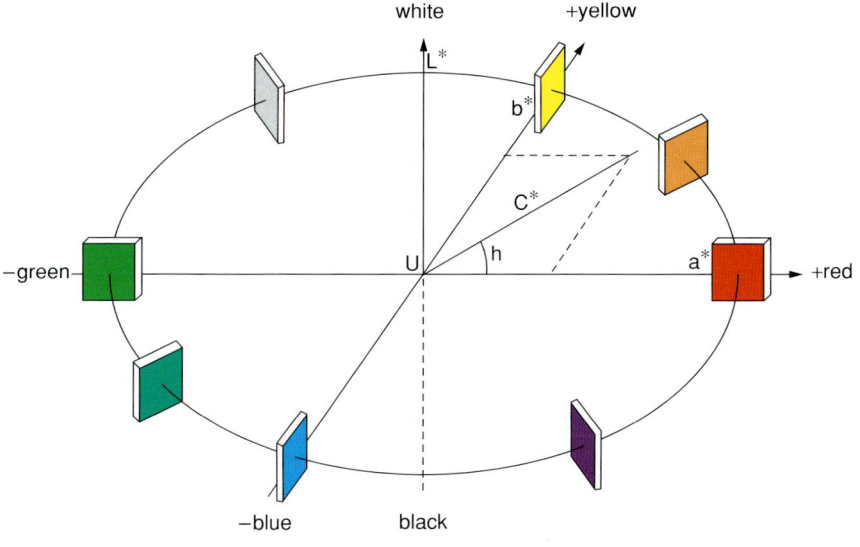

Color plate 1.

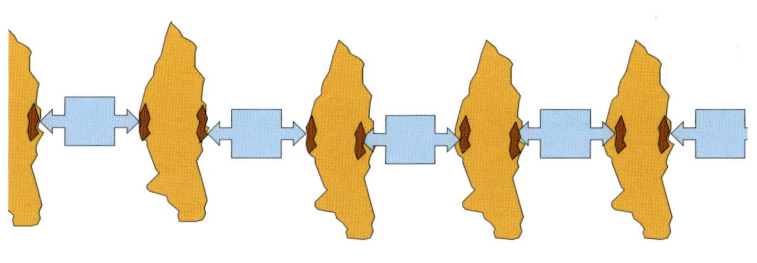

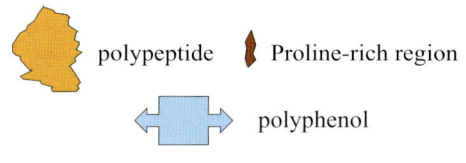

Color plate 2.

Haze

	6.	Oxidation during mashing leads to more removal of polyphenol precipitating with protein
	7.	Weaker worts contain more polyphenols
Boiling	1.	Vigorous rolling boil with high agitation precipitates haze-forming material
	2.	Use of carrageenan as precipitant
Hops/hop products	1.	Extracts contain less polyphenol
	2.	Using low-alpha hops for high-bitterness beers will introduce *pro rata* more polyphenol
Yeast and fermentation		Some haze-forming materials removed with yeast head
Conditioning		Ensure lowest temperature just short of freezing the beer
Filtration and stabilization	1.	Maintain beer as cold as possible through filter
	2.	Minimize oxygen uptake
	3.	Minimize pick-up of iron and copper
	4.	Kieselguhr superior to perlite. Latter may give "dull" product—one cause being dead bacteria originating in the malt
	5.	PVPP to remove polyphenols
	6.	Silica hydrogels/xerogels to adsorb haze-forming polypeptide
	7.	Tannic acid to precipitate haze-forming polypeptide
	8.	Papain to hydrolyze foaming polypeptide
	9.	Caution with addition of PGA foam stabilizer, which can precipitate with finings and papain
Packaging	1.	Lowest oxygen conditions
	2.	Avoidance of iron and copper pick-up
	3.	Avoidance of certain can lid lubricants
Final product	1.	Agitation during transport potentiates haze

|6|

Microbiology

Though it is remotely possible that some microbial spores from aerobic bacteria could survive kettle boiling of wort, the anaerobic conditions that follow would preclude their germination and growth. By all practical measures, therefore, brewers have the opportunity to start fermentation with a sterile raw material (wort). Microbial contamination therefore must enter from contact with air, from contaminated surfaces that wort contacts (e.g., pipes and tanks and pumps—see Chapter 14), via any insanitary additions to wort and from microbes residing in the pitching yeast. Because pitching yeast is recycled through the brewery 8 to 12 times (more or less), unwanted microbes *do* accumulate there, and so yeast is carefully monitored for contamination (see Chapter 11) and is regularly replaced by propagated yeast partly for this reason. Microbial contamination of beer is much less frequent than it once was, and so, these days, shelf-life is determined by loss of fresh beer flavor or possibly haze formation, rather than by damage caused by microbes. This is because (1) most beer is heat-treated in the final container or sterile-filtered before packaging and sale, (2) brewing equipment is well-designed and can withstand aggressive sanitation practices and (3) brewers have eliminated many traditional (and microbiologically rather risky) brewing practices. Complacency, however, will not do and so wise brewers remain vigilant.

The presence of contaminating microbes in brewing materials or beer is commonly described as "an infection" and the material is said to be "infected." This is an incorrect choice of words; only living tissue can become infected. The word "infected" implies a diseased state, disease transmission and the presence of pathogens (an incorrect and highly undesirable message for brewers to transmit). Nonliving matter becomes "contaminated"; thus, microbes in wort or beer are not an *infection* but a *contamination*.

Wort and, to a lesser extent, beer are good sources of nutrition for many microbes. Imposed upon this primacy, however, are the environmental conditions of brewing, particularly anaerobiosis and low temperature; together, these conditions prevent whole categories of microbes from spoiling wort and beer. There are some special conditions that influence the fortunes of contaminating microbes: (1) they face intense competition for nutrition from a very large population of yeast that favors its own success by using nutrients, lowering the pH towards 4.0 and producing alcohol(s) (to about 4%) all of which impede another wide swathe of potentially contaminating microbes and (2) wort contains iso-α-acids derived from hops α-acids (see Chapter 8) that in sufficient concentration are damaging to the membranes of sensitive microorganisms, especially at lower pH. Therefore, microbes that can cause serious damage to beer in process and to packaged beer are limited to relatively few kinds of bacteria and some "wild" yeasts.

The brewing environment is therefore an ecological niche for microbes, defined more by the *conditions* that pertain there than by the availability of nutrients. Therefore, processes in which the brewing conditions are less stringent are more at risk from microbes. These include, for example, open fermenters of small volume (air access); lower gravity worts and all-malt ones (this affects alcohol content and pH-drop); low hop rates (less toxic to microbes); higher fermentation temperature (encourages growth); and low pitching rate and/or sluggish yeast (limit competition). Furthermore, the introduction of an unusual brewing material (whether, e.g., unmalted grains or fruits, fruit infusions or syrups or spices) may affect the *conditions* of the process more than its *nutritional* aspects, with unexpected results. Of course, these materials might also introduce an unusual and large population of microbes, and so the process used must be such as to destroy them.

Contaminating microorganisms (1) cause unwanted off-flavors in beers even at quite low levels of contamination and (2) (of much less frequent concern) they can cause hazes, gels, slimes (rope) and pellicles that are difficult to remove. All organisms that spoil beer can tolerate the *conditions* they find there: low pH, high alcohol, no oxygen (low redox) and high CO_2, hop acids, coldness, low population; these are the inhibitory conditions, and they are more effective in combination than individually. The nutritional status of beer, while considerably lower than that of wort, is by no means exhausted, especially in the case of all-malt beers; it contains a useful range

Growth Media

Growth media can be "general" and employed for the enumeration of total levels of contaminating bacteria and yeasts, or "differential," for the purpose of counting specific types of contaminant.

Medium	Detects
Universal Beer Agar*	All types of organism including brewing yeast
Brewer's Tomato Juice*	All types of organism including brewing yeast
Wallerstein Laboratories Nutrient (WLN)*	All types of organism including brewing yeast
Lee's Multi-Differential	Bacteria
Raka Ray	Lactic acid bacteria
Lysine	Wild yeast
Lin's Wild Yeast Differential	Wild yeast
Barney–Miller	*Lactobacillus* and *Pediococcus*
DeMan Rogosa Sharpe (MRS)	*Lactobacillus* and *Pediococcus*
Malt Extract-Yeast Extract-Glucose-Peptone (MYGP) + copper	Wild yeast
Cadaverine, lysine, ethylamine and nitrate (CLEN)	Wild yeast
SMMP Medium	*Megasphaera* and *Pectinatus*

*General culture media. If nystatin is included in the media this suppresses the growth of yeast and allows detection solely of bacteria.

of nitrogenous materials including amino acids, sugar sources—particularly glucans (dextrins) and pentosans— vitamins and minerals for those organisms that can access them under the conditions that pertain. Fortunately, there are relatively few such organisms and none of them are pathogenic.

The organisms that spoil beer are usually divided into gram-positive and gram-negative kinds. This is not itself a useful division because the staining technique invented by Gram is rarely used in breweries. However, it so happens that gram-positive organisms (the lactic acid bacteria and "wild" yeasts) are common, frequent and important beer spoilers and spoilage by gram-negative ones is comparatively rare and limited, for the most part, to the early stages of fermentation. It is worth noting that many organisms listed in brewing *compendia* can be found in and around breweries, but by no means all of them actually spoil beer (i.e., enter and grow in beer and render it unfit for sale).

GRAM-POSITIVE BEER SPOILERS

The lactic acid bacteria are rods (some *Lactobacillus* spp) or cocci (some *Pediococcus* spp) that are sufficiently insensitive to hop acids to grow in beer. *Lactobacillus brevis* and *Lactobacillus casei* are among the most common rods encountered, and *Pediococcus damnosus* is by far the most important coccal spoiler of beer. In general, *Pediococcus* is more a problem of lager breweries, where its requirement for a low growth temperature favors it, and *Lactobacillus* of ale breweries, though this division is by no means absolute. *Lactobacillus* in sufficient numbers produces "silky waves" of turbidity that are quite unmistakable because the very long rods organize along the lines of sheer when a contaminated beer is rotated. In contrast, *Pediococcus* forms characteristic tetrads of spheres that are unmistakable when observed under the microscope.

Of central concern to brewers is that these organisms produce lactic acid to give beer a low pH and characteristic sourness. In addition, *Pediococcus* produces copious amounts of diacetyl (buttery or butterscotch flavor and aroma), though by a different pathway from that of yeast (see Chapter 11). Heterofermentative bacteria can make other products besides lactic acid, such as glycerol, ethanol and acetate, but these are of minor concern compared to lactic acid.

Control of contamination by lactic acid bacteria depends upon (1) rigorous sanitation practices (hygiene, see Chapter 14) in the brewery (2) regular replacement of the culture yeast with newly propagated yeast and, in many breweries (3) washing the cold yeast immediately before pitching with acid, most commonly phosphoric acid, for about 30 minutes at pH 2.2.

Direct examination of wort, beer or pitching yeast will not easily reveal the presence of lactic acid bacteria because the population of concern is too small, and many microscope fields must be examined to confirm the presence of the organism. Only high populations are easy to see, by which time the beer is unsaleable. Using plating techniques, detection of lactic acid bacteria in wort or beer samples is not difficult, but these classical methods are slow, taking a week or even more to give incontrovertible results; therefore, beers are usually released to the trade before the microbiology results are known. These results are historical but can be used to plot trends of processes to assure that they are "in control" (see Chapter 14).

The few lactic organisms that might be present in beer can be concentrated by passing, aseptically, a known volume of beer, say 100 ml, through a membrane (e.g., 0.2–0.45 microns), and then placing the membrane on the surface of a rich nutrient medium such as DeMan Rogosa Sharpe (MRS) (although there are many alternatives) designed to best support the growth of these organisms. Under anaerobic conditions (most effectively in a CO_2 atmosphere) and at about 20°C to 25°C, each viable cell present will soon

Pathways to Diacetyl

Figure 6.1 illustrates the pathway by which diacetyl is produced by yeast. Yeast can mop up the diacetyl, if it is healthy and remains in contact with the beer:

$$CH_3COCOCH_3 + NADH + H^+ \rightarrow CH_3CH(OH)COCH_3 + NAD^+$$
Diacetyl Acetoin

$$CH_3CH(OH)COCH_3 + NADH + H^+ \rightarrow CH_3CH(OH)CH(OH)CH_3 + NAD^+$$
Acetoin Butanediol

Yeast reductases reduce diacetyl successively to acetoin and 2,3-butanediol, both of which have much higher flavor thresholds than diacetyl.

Analogous reactions occur with pentanedione, the other significant vicinal diketone produced by yeast during fermentation.

Persistently high levels of diacetyl may indicate an infection by *Pediococcus* or *Lactobacillus*. Such bacteria produce much more diacetyl than pentanedione, and so if the ratio of diacetyl to pentanedione is disproportionately

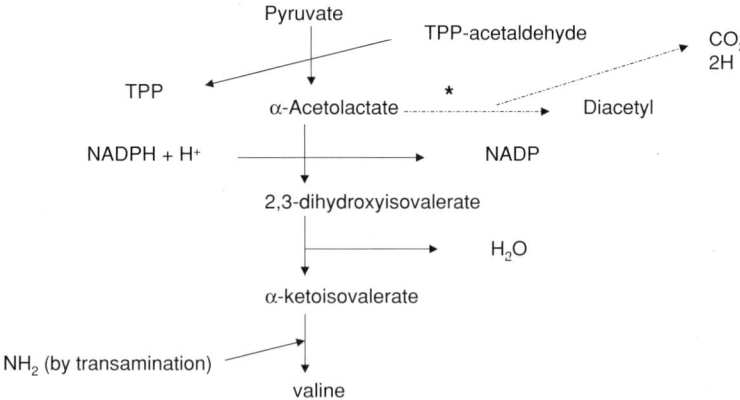

*Spontaneous oxidative decarboxylation (extracellular)

Figure 6.1. Diacetyl production as an offshoot of amino acid synthesis in yeast.

high, it suggests that there is an infection problem. The additional production by these lactic acid bacteria is due to an extra pathway by which diacetyl can be produced. This is via the reaction of the active form of acetaldehyde (thiamine pyrophosphate (TPP)-acetaldehyde, see Figure 6.1) with acetyl-CoA, with the formation of diacetyl and the release of TPP and coenzyme A.

---

yield a colony obvious to the naked eye or (more quickly) under magnification. The colonies can be counted. Note that incubation temperatures higher than 25°C or so will inhibit growth of *Pediococcus* to give reassuring, but wrong, results.

In contrast, detecting these unwanted microbes in the pitching yeast depends on suppressing yeast growth with cyclohexamide (actidione) or nystatin and often 2-phenyl-ethanol; these antibiotics are added to a supportive medium for lactic acid bacteria, e.g., MRS. The spread plates must be incubated anaerobically, whereupon colonies eventually arise. The bacteria that grow to form these colonies are obviously immune to the effects of the antibiotic. The identity of the contaminating organism can be confirmed by cell morphology (shape) under the microscope.

"Wild" yeast is often defined as any yeast in fermentation that is not the specific brewing strain required or intended. Most obviously, this suggests yeasts that are foreign to brewery fermentations, whether belonging to the genus *Saccharomyces* or another genus (see Chapter 11); however, an unintended *brewing* strain might also be considered "wild" as it might appear by cross-contamination in breweries that handle more than one brewing strain. Brewers in such breweries take stringent precautions to prevent cross-contamination because it can be difficult to detect and to eliminate. Contamination of brewery fermentations by "wild" yeast(s) possibly occurs much more frequently than we are aware of because some contaminants behave so similarly to the culture yeast that there is no real consequence to their presence and so no easy way to detect them; alternatively, a wild yeast contaminant could behave *so* differently from a brewing strain that it is quickly out-grown by the brewing strain and so eliminated. We are much more cognizant of "wild" yeasts whose presence is signaled in some symptomatic way, e.g., by change of flavor or pH, or growth rate, or fermentation rate, yeast crop, or early or late flocculation, or the end gravity (high or low) achieved, or, in the very rare case of a "wild" yeast that can produce a "killer factor" (zymocide) that causes the elimination of the brewing strain. The off-flavors associated with "wild" yeast generally fall within the normal spectrum of yeast flavor compounds; that is, wild yeasts tend to accentuate estery, or alcoholic or acidic or sulfury flavors, or produce excess diacetyl.

4-Vinylguaiacol

When ferulic acid is decarboxylated by the enzyme ferulic acid decarboxylase, the product is 4-VG, which has the distinct aroma of cloves. The enzyme is not present in lager strains and most ale strains of yeast but is present in wild yeasts, so the presence of 4-VG in most beers is a sure sign of contamination by wild yeast.

The only brewing strains that elaborate this enzyme are those ale yeast employed in the production of the Bavarian weizenbiers. Accordingly, a clove note is very much a sign of authenticity in such beers and is not considered a flavor defect in the style.

However, the classical off-flavor associated with "wild" yeast is medicinal, or phenolic, or bandage, or clove-like, and arises from the decarboxylation of wort phenolic acids (e.g., ferulic acid) to produce compounds like 4-vinylguaiacol (4-VG). This quality, however, is an essential character of some beers, particularly traditional wheat beers.

Control of wild yeasts in the brewery depends on rigorous sanitation practices and regular yeast replacement. Acid washing is ineffectual for removing "wild" yeast.

Wild yeasts are generally divided into those that belong to the genus *Saccharomyces* and those that belong to other genera. We generally assume that *Saccharomyces* "wild" yeasts are more akin to brewing strains and hence less likely to be sharply distinguished from brewing strains by the following tests than are non-*Saccharomyces* yeasts; however, no test is absolute. Success or failure depends on the particular mix of yeasts in a pitching yeast population and the particular properties of the brewing yeast itself. Most breweries use several tests and adapt them to their own particular circumstance. Detection of wild yeast might be as simple as direct microscopic examination. The 37 genera of yeasts include yeasts with a rich variety of shape and size. Many wild yeasts, including nonbrewing *Saccharomyces* yeasts, are simply smaller or more shapely than their large, rotund and portly brewing yeast cousins and, if present in sufficient numbers (i.e., percentage of total cells), are easily detected by direct observation. Wild yeasts generally grow more readily under adverse circumstances than brewing strains; thus, a medium with the amino acid lysine as the sole major source of nitrogen favors most wild yeasts. Similarly, most brewing yeasts are susceptible to inhibition by copper or crystal violet and so those yeasts growing in media supplemented, to a sufficient level, with copper or crystal violet (or both) are probably "wild." Again, wild yeasts are generally less sensitive to the inhibitory effects of cyclohexamide (actidione) and will grow on media supplemented, to a suitable low level (about 5 mg/kg), with this antibiotic. The colonies of different yeasts take up the pH indicator bromocresol green in characteristic ways and can sometimes even distinguish one brewing yeast from another. A mixed yeast population spread on an actidione plate may comprise white colonies (wild yeast) and green ones (brewing strains), although the picture is rarely so clear cut; also, wild yeasts tend to yield colonies that are small, relative to brewing yeasts, under the same growth conditions on nutrient plates. Contamination of lager yeasts in an ale yeast population can be detected by spread-plating a yeast sample on a melibiose-containing medium; lager yeasts contain melibiase and can grow to form colonies. Ale yeast contaminating a lager yeast population can be detected by incubating a spread plate at a temperature above the maximum temperature of growth for lager yeasts (about 32°C). The ale yeast will grow to form colonies.

GRAM-NEGATIVE BEER SPOILERS

All the gram-negative organisms that spoil beer exploit a breakdown in the inhibitory *conditions*, listed above, that normally protect beer. For example, the acetic acid-producing bacteria (*Acetobacter* spp and *Gluconobacter* spp) are *aerobes* and tend to be found, e.g., around spilled beer, around poorly maintained bar taps and as contaminants of cask beer to which air has gained access. The terrifying *Zymomonas* spp that tolerates ethanol and produces noxious quantities of H_2S and acetaldehyde (as well as alcohol) can use only glucose or fructose and requires quite warm conditions; therefore, it appears mainly in primed cask-conditioned ales. *Megasphera* spp (cocci) and *Pectinatus* spp (rods) tend to be intolerant of alcohol and low pH and therefore can spoil only weak beers, contributing fatty acids and H_2S characters. The Enterobacteriaceae (short rods, some motile) that are related to beer processing include *Obesumbacterium, Enterobacter, Citrobacter* and *Klebsiella*; all these organisms tend to be intolerant of alcohol and low pH and so are primarily wort spoilage organisms; they enter during wort cooling or by insanitation; they are not beer spoilers, and fail to survive fermentation. During fermentation, if present in sufficient numbers for sufficient time, this group of organisms can make a wide range of flavor-active compounds, typically described as cooked-corn/vegetable/faecal (mostly sulfur compounds) to sweetish/fruity aromas. *Obesumbacterium* may be an exception: in ale fermentations it survives fermentation and separates with the yeast and is therefore recycled to the next fermentation. However, even in

Rapid Microbiological Methodology

Traditional methods for assessing the extent of contamination of beer and brewery process streams are time-consuming, involving the introduction of swabs or liquid samples to solid media in dishes, incubating typically for three days (aerobic organisms) and seven days (anaerobic organisms) and then counting the plates.

Nowadays, a series of more rapid procedures is available. They may involve a pre-concentration stage, such as capture of organisms on a sterile membrane filter with pore-sizes of 0.22 or 0.45 μm. In the direct epifluorescent filter technique (DEFT), the cells trapped on the filter are exposed by staining them with a fluorescent dye such as acridine orange. Viable cells stain orange and dead cells stain green.

Perhaps the most widely used rapid microbiological technique is the ATP bioluminescence test. Wherever there are organisms, alive or dead, ATP is to be found. The kit employed contains a firefly enzyme, luciferase, which reacts ATP with a substrate called luciferin to produce light. There is a proportionality between the intensity of light produced and the amount of ATP present, and hence the extent of contamination.

---

this case, replacing the yeast on a regular basis and yeast washing (neither of which is a condition of traditional ale-brewing practice) would control this contaminating organism.

Control of these organisms depends again on simple but sound brewing practices, including scrupulous sanitary practices in the brewery and replacement of yeast on a regular basis. Acid washing of yeast effectively eliminates them.

Most beers receive additional treatment to assure microbial stability in the marketplace. Sufficient heat treatment to reduce a suitable test population of microorganisms by 12 log cycles, by convention, sterilizes any substrate. Such extensive heating severely damages the flavor of many foods, including beer. Pasteurization is not sterilization; it is sufficient heat treatment to reduce a test population of microbes to an approximately known and satisfactory level of risk. Successful pasteurization of beer therefore depends critically on upstream processing, including the nature of the microbial population present (especially *Lactobacillus* spp), the number of cells present, the nature of the beer, the size of the package and the level of microbiological reassurance the brewer needs. Fortunately, beer contains vegetative

Sterile Filtration or Pasteurization?

Some brewers are adamant that the pasteurization of beer damages its flavor, and therefore prefer to remove microorganisms by so-called sterile filtration. In this process, beer after mainstream filtration and stabilization is filtered again, most significantly by passage through filters with pore-sizes below 0.45 μm. Sometimes brewers market their beers on a platform of "cold filtered," which is of course ludicrous when most beers are filtered cold, whether pasteurized or not.

Many brewers find that, provided the beer contains very low levels of oxygen (e.g., < 0.1 ppm), it will comfortably tolerate pasteurization (< 20– < 30 PU) without any impact on flavor. Those that do see an effect may be the brewers where weaknesses in the process have led to there being high levels of precursors molecules in the beer which, when heated, degrade to release aged characters.

cells not spores (spores are very heat resistant) and a low population of contaminating cells (that are relatively easily eliminated); beer is acidic and alcoholic (and so an unsupportive environment for survival of microbes) and the packages are usually quite small (so heat penetrates quite rapidly). Relatively little heat treatment is therefore necessary.

One pasteurization unit (PU) is 1 minute at 60°C (140°F). For a test population of microbes relevant to brewing, a 7°C (12.5°F) change in pasteurization temperature yields a tenfold change in delivered PUs or, alternatively, a tenfold change in the time of heat exposure. Thus, 53°C for 10 minutes and 67°C for 0.1 minutes both deliver 1.0 PU. For beer, this can be conveniently expressed as $PU = 1.393^{(T-60)}$. Most brewers use 5.0 to 15.0 PUs (delivered at the center of the bottle or can), though much higher levels are used in some cases, depending on the particular circumstances. As implied above, least heat is required for beers that are low in pH, high in alcohol, are brilliantly clear and have very low microbial counts, and whose shelf-life is controlled.

Beer in the final sealed package (which is the great advantage of this technique) can be heated in a tunnel pasteurizer; the packages pass in an endless stream through a large tunnel in which the beer is heated, held and cooled by water sprays. The disadvantages are as follows: large complex equipment, relative inefficiency of heating/cooling, relatively high water use, exposure of *packaged* beer to heat, overpasteurization of packages should the

High Pressure

Most organisms are sensitive to increased pressure and there has been the exploration of high-pressure technology to achieve microbe kill in several foodstuffs. Although explored for beer, with pressures of 300 MPa proving effective, there is no commercial practice of this type in breweries as yet.

conveyer stop, wet bottles/cans for labeling, "bloom" or water spots on bottles, steam in the packaging hall, etc. The alternative method of "bulk" pasteurization, in which *cellar* beer passes in a thin continuous stream through a heat exchanger, is a much preferred technology in all these aspects; also, brewers often use an HTST (high-temperature-short-time) strategy of heating in such equipment, so called "flash" pasteurization, in which the beer is heated to a high temperature for a very short time; this minimizes heat damage to the product. The one overriding disadvantage of "bulk" or "flash" pasteurization is that the pasteurized beer must now be packaged aseptically. This is a daunting technical challenge and is undoubtedly the reason that tunnel pasteurization remains an extant technology.

Aseptic packaging is also required for beers that are freed of microbes by filtration, using membranes, very tight sheets or ceramic filters; the one additional advantage here is that beer in not exposed to heat at all.

Beer Design for Increased Resistance to Spoilage

Susceptibility to spoilage will be less if

(1) bitterness is high
(2) pH is at the lower end of the 4.0 to 4.5 range
(3) free amino nitrogen level tends towards zero
(4) ethanol content is high
(5) oxygen content is low
(6) CO_2 content is high
(7) carbohydrate level is low.

|7|

Inorganic Ions

Inorganic cations and anions in beer arise from the brewing raw materials, especially malt, and from water and specific additions of salts that brewers might choose to make. Traces of ions might derive from the brewing plant itself. The roles that these ions play in successful beer-making is often subtle and indirect, but it is not trivial; in some places and on some occasions inorganic ions are a defining character of the beer produced.

The inorganic composition of barley, and the malt made from it, reflect to some extent the soil upon which it was grown and the soil treatment. The most common ions in malt are therefore potassium and phosphate and these are readily extracted into wort during mashing. Phosphate is primarily bound up in phytin, a component of the living tissue of the barley, and is mobilized by hydrolysis during germination. Modification (see Chapter 9) therefore increases the availability of inorganic phosphate. If phosphatases survive kilning (e.g. in pale lager malt) and the mash is initiated at low temperature, then phytin can also be hydrolyzed in mashing to yield the inorganic ion. There is a good deal of phosphate in wort (about 600–800 mg/l); roughly half of this survives into beer where it is a benign presence without any known direct effects on beer qualities (flavor, foam, clarity and the like). However, phosphate (and phytin) has a major role to play in establishing the pH of the mash, wort and beer (see Chapter 2) and in this indirect sense is pivotal. Particularly, these ions can react with Ca^{2+} and, to

Phytin, Phosphate, Calcium and pH

Through the interaction of calcium and phosphate, the pH in brewing systems can be lowered, according to the equation

$$3Ca^{2+} + 2HPO_4^{2-} \rightarrow Ca_3(PO_4)_2 + 2H^+.$$

Hard water can cause mash pHs up to 0.3 units lower than those made with soft water. A rise in calcium level from 50 to 350 ppm drops wort pH from 5.5 to 5.1. Calcium also reacts with carbonate and polypeptides to promote the release of protons.

Phosphatase enzymes called phytases in malt attack phytic acid (inositol phosphate or phytin, a storage form for phosphate in grain) to release phosphate. Ironically, lowering the pH by addition of acid increases the activity of phytase, this acts to release phosphate which, by reacting with calcium, further presages a pH drop.

a lesser extent, with Mg^{2+} to form insoluble salts and so cause acidification (lowering of wort pH). This is certainly one cause for the loss of phosphate from wort in the brewhouse e.g. during boiling, when Ca^{2+} is often added to wort as Burton salts (mainly gypsum, $CaSO_4$) or as $CaCl_2$. Phosphate is also necessary for yeast nutrition and is taken up by the yeast. However, it is difficult to imagine that fermentation efficiency would ever be compromised by lack of phosphate in the wort. Phosphate can also enter wort with the pitching yeast because phosphoric acid is commonly used for acid-washing of yeast (see Chapter 11).

Potassium in wort and beer approaches 500 mg/l and is derived primarily from the malt though it could be added as KMS in some beers; thus wort gravity and adjunct ratio (especially if syrups are used, being particularly low in inorganic ions) affect the concentration of this ion. Because chloride ion is also a common component of wort, potassium ion (as KCl, along with NaCl if present in wort naturally or if added, e.g. at the kettle boil), can add salty or mineral characters that contribute to the perception of "body" or palate fullness. Na^+ and Cl^- also arise primarily from the malt.

Malt, and to a lesser extent other raw materials (e.g. water), also contributes to wort a broad spectrum of elements, useful in traces in yeast nutrition, most of which survive in very small, but measurable, quantities

Speciation

It is too simplistic to discuss metal ions as if they were present in brewing systems in pure solution. They can enter into complex formation with a number of other substances. One example of this is "chelation": this is a reaction between molecules that through charge–charge interaction can bind onto metal ions, whilst remaining in solution. Molecules such as amino acids, peptides, polypeptides and organic acids have this ability. Alternatively, the metal ions may bind onto relatively insoluble substances and be removed from solution. For example this may occur with trub and in the formation of haze.

The speciation is significant in respect of an ion's ability to exert its impact. For instance, free copper ions can be potent oxidants if they are able to react with oxygen to activate it. If the copper is attached to a chelating agent, however, it may no longer be so damaging. (Alternatively, sometime the chelated form is more potent.)

into beer. These include copper, iron, manganese and zinc. Copper (and possibly iron) can also enter wort from wort/beer contact surfaces; many brewers think that a strictly controlled small amount of copper has a desirable role to play in beer-making—especially in the control of sulfury aromas

The Role of Zinc

Zinc is important for the action of several enzymes, notably alcohol dehydrogenase and superoxide dismutase. It is probably also important in the regulation of gene expression by entering into complexes with peptides that can interact with DNA.

Zinc can also serve to bridge between iso-α-acid residues within the complexes with amphipathic polypeptides that occur in the bubbles of foam. To effect a significant increase in beer foam stability zinc needs to be present in beer at levels some ten times higher than is routinely added to promote fermentation.

during fermentation. Iron can enter beer also from filter aids especially from "beer soluble iron" in diatomaceous earth (a component for which there is a standard ASBC analytical method). Copper and iron, can participate in oxidations and reductions and accelerate loss of fresh beer flavor; these ions also react vigorously with proteins and concentrate strongly in haze materials and in beer foam and contribute to their formation (see Chapters 4 and 5). In any significant amount these ions are toxic and mutagenic to yeast. Zinc is commonly added to wort (say 0.2 mg/l) at the kettle as a yeast nutrient because it is often deficient in wort. Zinc is a component of several key enzymes including yeast alcohol dehydrogenase (ethanol from acetaldehyde) and aldolase (of the EMP). Also, zinc at about 0.2 mg/l improves yeast vitality and survival.

Malt also contributes to wort magnesium, sulfate, calcium and chloride ions but it is usual to consider these ions as arising from brewing water. The divalent ions can be a component of the brewing water as it enters the brewery, or the level(s) of these ion(s) might be significantly boosted by adding salts to the water, mash or kettle, as "Burton salts" (mostly calcium with magnesium present primarily as the sulfates), as gypsum (essentially calcium sulphate) or as calcium chloride. The last is commonly used because it dissolves easily and the chloride ion (with K^+ and Na^+ undoubtedly) has a positive effect on perception of mellowness and palate-fullness ("body"); it also helps to counteract the dry, bitter and even sour character of excessive sulfate in some beers (especially as $MgSO_4$ or Na_2SO_4). The divalent ions Ca^{2+} Mg^{2+} play an essential part in the pH of brewing systems and so affect many aspects of brewing technology (as mentioned above and fully explored at Chapters 2 and 13 and not repeated here). In addition to these pH-based effects, calcium helps to precipitate oxalic acid as calcium oxalate in the brewery, which otherwise would appear in beer and might cause gushing. Calcium is also intimately involved with the mechanism of yeast flocculation (see Chapter 11). It stabilizes malt α-amylase during mashing. Magnesium by contrast has much less influence on pH than Ca^{2+} because it is generally present in much lower concentration (e.g. 25 mg/l vs 250 mg/l) and also the magnesium salts of phosphate or phytin are much more soluble than those with calcium. Magnesium, however, is a component of essential yeast enzymes such as pyruvate decarboxylase and all kinases (ADP/ATP-handling enzymes).

Water uniquely introduces into brewing only one ion: that is the bicarbonate (HCO_3^-) ion. Bicarbonate is present in most waters, especially ground waters, although the amount can vary over a wide range. It can be present in "soft" water e.g. as $NaHCO_3$ or as the "temporarily hard" component of hard water $Ca(HCO_3)_2$. In either case, upon heating, the salt decomposes to yield the hydroxyl ion (OH^-) and high alkalinity. There is no virtue in alkalinity for brewers and this ion is therefore strictly controlled in a variety of ways (see Chapter 13).

Silica

Beer can contain silica, which originates in the outer layers of grain and from the water supply. It can cause scale in steam generation systems in the brewhouse, through interactions with calcium and magnesium. It can also contribute to hazes. However, it may also have beneficial effects: if present in beer it can make a contribution to bones of consumers, as well as chelating undesirable ions, including aluminum.

III

Processes

| 8 |

Raw Materials

The approach of this book to its subject matter is longitudinal; that is we try to look down the length of the brewing process to detect how beer properties arise and how, at multiple stages, the outcome of the process is affected. The book could have therefore been written with but two chapters entitled (a) Raw Materials and (b) Processes. Much of what we might want to say about raw materials, however, is subsumed in other chapters, e.g., Modification, Color, Enzymes and pH, to satisfy the structural plan of the book. There is intentionally therefore not a great deal of material left over for a long chapter on raw materials. The main motivation for including such a chapter at all is that we have made relatively little mention of hops and that is curious, if not fatal, in a brewing text; that is where we begin this chapter.

HOPS

Hops are added in brewing in either or both of two places: in the kettle and/or after fermentation. The objective is the same in each case: to make beer bitter to an exact, consistent and repeatable level. Over the lifetime of most senior brewers in the industry, no raw material has evolved and improved more than hop products. The starting place for hop products

Chapter 8

> ### Assessing Resins in Hops
>
> In the United States, hop α-acids are measured spectrophotometrically. They are extracted with toluene and the absorbance of the extract after dilution in methanol is assessed at 275, 325 and 355 nm.
>
> $$\alpha\text{-acid}(\%) = d\,(-51.56 A_{355} + 73.79 A_{325} - 19.07 A_{275})$$
>
> where d is a dilution factor (typically 0.667).
>
> As hops age and deteriorate the ratio of A_{275} to A_{325} increases.
>
> In some locations, α-acids are assessed from a titration with lead acetate and monitoring of electrical conductivity. As increasing quantities of lead acetate are added, there is a decrease in conductivity up to the point at which there is an equivalence in concentration of the positively charged lead and negatively charged hop resins, after which the conductivity increases again. The more lead acetate needed to reach the "low point," the more resin in the hops.

is where it has always been—in the hop yards of Germany and Europe generally and some western states of the USA; there are increasing and improving supplies from China and Australasia. Hops are prized eponymously as aroma hops or bittering hops. Their uncompromised quality is important to brewers. Beyond such expected analyses as moisture, bittering potential (α-acid content) and the hop storage index (A_{275}/A_{325}), brewers are concerned with contaminants, especially pesticide residues. Dried compressed whole hops or hop cones (sometimes incorrectly called "leaf" hops or hop "flowers") are then converted to a range of hop products, such as pellets or extracts, that are more stable to oxidation, more compact, more easily transported and stored, more standardized (e.g., in terms of α-acid content and blend of varieties), easier to quantify and easier to use in practice; e.g., with automatic addition and with less (or zero) waste that is more easily handled, and some extracts that are insensitive to light. As a result, in the long run, these products are more cost effective than whole hops, and most beers these days are made with some form of hop pellet or hop extract or combination of the two.

Hops add bitterness to beer and this is a primary reason for using them, as beer without any bitterness is bland and satiating. Nevertheless, bitterness is not a quality most consumers like and so hop bitterness must be artfully melded with the other qualities of the beer. Commonly, therefore, there is a good correlation between the overall flavor impact of a beer derived from malt and yeast and its analytical bitterness; however, beer pH also affects the perception of bitterness because it affects the dissociation of the iso-α-acids present. Lower pH generally yields finer but less intense bitterness as the α-acids associate. For the same reason, lower beer pH also increases the antimicrobial properties of iso-α-acids, provided they are present in sufficient amount. Hop compounds also help to stabilize beer foam, which, in the case of compounds like tetra-hydro-iso-α-acid, is a particularly notable and characteristic quality. Iso-α-acids are hydrophobic and tend to separate onto surfaces of all kinds, e.g., vessels, trub, yeast; this accounts substantially for the relatively poor utilization of α-acids added to the kettle. Iso-α-acids migrate into foam bubbles where they might well form complexes with, e.g., proteins that are detergent like, surface active and hence foam stabilizing. Beer foam is always more bitter than the beer from which it came.

α−acids of hops are insoluble in beer and so for all practical purposes they are not bittering. When heated near boiling, however, they melt and isomerize (change their molecular shape) to iso-α-acids, and in a vigorous "full rolling" boil these are reasonably soluble in wort and persist into beer. A full rolling kettle boil is also necessary for wort sterilization (see Chapter 6) and precipitation of some unwanted protein (see Chapters 4 and 5). Higher wort pH and the presence of ions, especially Ca^{++}, accelerates isomerization reactions and doubtless promote higher utilization of hop α-acid; however, desirable *quality* of bitter flavor and wort color are compromised at higher pH and hot break is reduced. This might not be of particular concern for dark beers and stouts but is undesirable in pale and delicate ones. Utilization (%) of hop substances in the wort kettle is reduced by high wort gravity, especially in all-malt worts, and by high hopping rates; generally, utilization is affected by the tendency of iso-α-acids to attach to surfaces of vessels (hence small vessels reduce utilization) and to "break" or "trub" particles, to finings and filter aids and to yeast cells and fob during fermentation. Thus, the kettle efficiency is only a part of the overall efficiency of hop utilization.

The need to improve hop utilization (from, e.g., 40% to 85%) has driven a significant move to postfermentation hopping, using isomerized hop extracts of various kinds. These days, they are made mainly from hop α-acids extracted with CO_2 and then treated with divalent metal ions (usually Mg^{2+}) under slightly alkaline conditions. This technology has also permitted light-stable hop extracts to be produced by reduction of the susceptible double bond of the iso-α-acid molecule; this prevents the formation of 3-methyl-2-but-1-ene thiol (iso pentenyl mercaptan) that has the aroma of skunks.

Hop Utilization

Hop utilization is an expression that quantifies the extent to which the resins available in hops and hop products manifest themselves as bitterness in beer:

$$\text{Hop utilization}(\%) = \frac{\text{iso-}\alpha\text{-acids in beer} \times 100}{\alpha\text{-acids introduced}}$$

For cone hops the value is usually rather low, e.g., 25% to 30%. The brewer will therefore need to add proportionately more hops to the kettle to account for the poor efficiency of conversion. Pellets and extracts added to the kettle will afford better utilization, but naturally the best values are obtained when isomerized extracts are added to the finished see and there is the least opportunity for the relatively insoluble molecules to be lost by adhesion to yeast and other particles. Note that in the strictest sense of the term, utilization for a isomerized material involves comparison of iso-α-acids remaining in beer with iso-α-acids, rather than α-acids added. Another key advantage is that the use of such materials means that less undesirable material is added. If, for example, a cone of whole hop addition was made from a hop of relatively low α-acid content with poor utilization, then to achieve the desired bitterness in a product of significant BU value would mean that considerable vegetable matter would have to be added to the kettle. Within that material would be substances such as polyphenols that may be undesirable, e.g., from a colloidal stability perspective.

BARLEY

Barley, almost always in the form of malt, provides the bulk of the extract for most worts, and is an essential source of nonsugar nutrition for yeast comprising amino acids, vitamins and minerals. For making malt, barley must be of a suitable malting variety, sufficiently low in protein (11%–13% as N × 6.25), adequately free of dockage and skinned and broken materials, highly viable (at least 96%) and quite low in moisture (12%–14%), and the lot should have a high proportion of plump grains. Ideally, the endosperm should be "mealy" (i.e., paper-white and opaque and containing a myriad of

Hop Products

Product	Nature	Stage of addition
Leaf hops	Whole hop cones	Kettle
Pellets	Cones hammer-milled, blended to desired mix, extruded	Kettle
Isomerized pellets	As for pellets, but isomerization agent added to powder and pellets held warm to promote isomerization	Kettle
Hop extracts	Liquid carbon dioxide extraction of powdered hops	Kettle
Isomerized kettle extract	As for hop extract, but extract isomerized	Kettle
Isomerized extract	Hop extract fractionated into resin and oil components and resin isomerized	Beer
Reduced isomerized extract	Isomerized extract with addition of two (rho), four (tetra) or six (hexa) hydrogen atoms to afford increased light resistance	Beer
Aroma extract	Hop extract fractionated to remove resin component	Beer
Late hop extract	Aroma component from extract separated into spicy and floral components	Beer

air-cells throughout the endosperm) not "steely" (i.e., somewhat grayish and translucent—a condition promoted by high protein, poor growth conditions and inadequate storage). Mealy grains take up water more rapidly and evenly during steeping.

Barley is harvested in the fall (in the northern hemisphere), but is malted all year and so must be stored. Successful storage depends upon grain temperature, moisture content of the kernels and time as well as conditions like the purity and cleanliness of the lot and absence of broken grains. The enemies of barley in storage are microbes, insects and grain respiration and neither dryness nor coolness protects them from all the enemies; the grain must be stored cool *and* dry, e.g., a rule-of-thumb "13/13," that is no more than 13°C and 13% moisture. The grain must be moved and cleaned on a regular basis. Prolonged dry storage permits the grain to pass through dormancy and water sensitivity (most easily construed as residual dormancy) until it is ready for malting. The last of a season's crop is usually malted some 15 months or so after harvest, and this is a typical target storage period.

After the barley is selected, stored, separated (to remove unwanted grains and broken kernels) and graded (to select kernels of the same size), it is brought to the barley washer and thence to the steep tank that initiates the malting process; the grain takes up water and swells and life processes resume. Respiration (an oxygen consuming process) rises and throughout the steeping process maltsters provide adequate aeration to prevent stifling of the grain. Over about 48 hours, the moisture content of barley rises from about 12% to a target moisture content in the range 42% to 48% depending on the maltster's objective and the characteristics of the barley. Generally, high steep-out moisture is used to make colored malt or to achieve high modification (at the cost of high malting loss) or if the barley is slow to germinate for some reason. Pale malt is generally made from vigorous barley, and the steep-out moisture is therefore at the low end of the range. If gibberellic acid (GA_3, with or without bromate) is used to promote germination, it is usually added to the final steep water or to barley en route to the germination vessel. The level used is about 1 mg/kg of dry barley.

After steeping, barley is drained and moved to a germination vessel that is designed to make it easy to control the conditions under which the barley germinates. This in turn controls the modification of the barley endosperm, i.e., the sum of changes that barley undergoes as it is transformed from barley to malt. These changes are so important and far reaching that we devote a separate chapter to modification and its effects in brewing (see Chapter 9). Germination takes about four days, during which time the temperature of the grain bed rises from about 15°C to about 20°C, despite (1) constant application of a stream of cool humid air throughout the period of germination and (2) regular turning of the grain to promote even air flow and prevent entanglement of rootlets. As the grains grows during germination, it breaks down its own storage substance (the endosperm materials) to provide energy and matter for embryo growth; this causes heating up of the grain bed and malting loss, i.e., the loss of dry substance as CO_2 and H_2O are formed during ATP generation. Available extract is additionally lost as rootlet (counted as part of the malting loss) and shoot. During germination, barley is capable of producing from the aleurone layer and scutellum, a sufficient spectrum of enzymes to reduce the endosperm substance entirely to its basic building blocks—e.g., primarily sugars and amino acids—and given time could consume them entirely to build the substance of a new plant. This occurs in the field but is assiduously avoided by the maltster who curtails germination at the most opportune moment

Enzyme action in malting and brewing

Process Stage	Treatments	Events	Endogenous Enzymology	Exogenous Enzymology
Raw barley	Storage—perhaps to break dormancy	Hormonal changes—ill-defined	Few enzymes in raw barley: Main ones carboxypeptidase and bound, inactive β-amylase	
Steeping	Water added, interspersed by air rests, to raise water content of embryo and endosperm; up to 48h at 14–18°C	Synthesis of hormones by embryo, hydration of 'substrate' (starchy endosperm)	No apparent increases reported	
Germination	Controlled sprouting ('modification') of grain—typically 4–5 days at 16–20°C	Synthesis of enzymes by aleurone and migration into starchy endosperm; sequential degradation of cell walls, some protein, small starch granules and pitting of large granules	Solubilization of β-glucan by solubilases (?) and endo-β-glucanase; degradation of arabinoxylans by arabinofuranosidase and endo-xylanase; partial hydrolysis of proteins by endo-peptidases and carboxypeptidase; development and limited action of α-amylase; splitting of β-amylase from protein Z; synthesis of bound and free limit dextrinase, and activation of latter	Microbial flora may contribute enzymes; opportunities for use of selected starter cultures?

Process Stage	Treatments	Events	Endogenous Enzymology	Exogenous Enzymology
Kilning	Heating of grain through increasing temperature regime (50–220°C) for desired properties: enzyme survival, removal of moisture for stabilization, removal of 'raw' flavors, development of 'malty' flavors and color	Enzyme survival greater with low temperature start to kilning and lower final 'curing' temperature. Increased heating of malts of increased modification (i.e. higher sugar and amino acid levels) gives increasingly complex flavors and colors via Maillard reactions	Some continued action of all enzymes at lower onset temperatures; but then solely an enzyme inactivation issue. Lability of enzymes endo-β-glucanase, limit dextrinase, lipoxygenase> endo-peptidase> β-amylase, lipase> solubilase > α-amylase, peroxidase	
Malt storage	3–4 weeks ambient storage, otherwise wort separation problems later	Unknown, but may relate to development of cross-links between proteins through oxidation in mashed of unstored malt	Lipoxygenase may catalyze this reaction (c.f. enhancement of bread-making by analogous reaction in wheat protein); lipoxygenase decays during malt storage	
Mashing	Extraction of milled malt at temperatures between 40 and 75°C	Enzymolysis continued; gelatinization of starch at >62°C	Continued β-glucanolysis favored at low temperatures—also possibly further proteolysis; starch degradation greatly facilitated by gelatinization; balance of enzymes acting faster at higher temperatures with increased destruction of more sensitive ones	Use of heat-stable β-glucanase from Bacillus or fungi comprises main use of exogenous enzymes in high malt mashes; use of glucoamylase to promote fermentability (Light beers)
Use of adjuncts	Solid adjuncts used in brewhouse, taking advantage of malt enzymes (liquid sugars are products of acid and enzyme action in sugar factory and added at boiling stage)	Cereals with higher starch gelatinization temperatures than for barley need pre-cooking before combining with main mash	Ditto—also a degree of dilution of malt enzymes, especially with high adjunct use	Use of highly heat-resistant α-amylase to promote gelatinization in cooker. Use of amylase, protease, β-glucanase mixtures in main mash

Stage	Conditions	Purpose	Biochemistry	Comments
Boiling	1–2h at 100°C, before cooling	To sterilize, extract hops, concentrate, and kill all residual enzymes	No enzymology	
Fermentation	Wort pitched with yeast and fermented for 3–14 days at 6–25°C	Fermentation of glucose, maltose, sucrose, maltotriose to alcohol; enzymic production of various flavorsome compounds (alcohols, esters, fatty acids, sulfur-containing compounds etc) Synthesis and removal of diacetyl as an offshoot of amino acid production	Embden-Meyerhof-Parnas pathway and offshoots	Addition of acetolactate decarboxylase to convert acetolactate precursor to acetoin, thereby circumventing 'natural' route which is non-enzymic breakdown of acetolactate to diacetyl (butterscotch), which is slowly reduced by yeast enzymes to less flavor-active acetoin
Cold conditioning and filtration	−1°C for ≥ 3 days; then filtration	Precipitation, sedimentation and removal of solids	Slow action by any enzymes 'leaked' from yeast, e.g. proteinases: detrimental to foam	Filtration can be limited by viscous polysaccharides, ergo advantage of using β-glucanase in brewhouse (or fermenter). Some use papain as a haze-preventative—but risk of removing foam polypeptides
Package	Market-driven	Progressive deterioration by chemical reactions, including oxidation	Unpasteurized, 'sterile-filtered' beer, may retain some of these enzymes	Use of glucose oxidase/catalase as an oxygen scavenger has been suggested

for malt quality by achieving sufficient modification with minimum malting loss.

Enzymes are proteins (see Chapter 10) and therefore are prone to denaturation by heat. During kilning of malt profligate enzyme destruction does occur and the enzymic quality of dry malt is a shadow of the green malt from which it is made, both in terms of the *amount* and *kinds* of enzymes present; only these surviving enzymes are carried forward into mashing in the brewery. Although traces of many enzymes might survive kilning, brewers evaluate malt on the presence of only the starch-digesting amylases: they measure DP or diastatic power (also DU or dextrining units) (see Chapters 9 and 10). From the point of view of wort quality, it is best to assume that the primary action of enzymes, other than amylases, is confined to the malting process and that their action in mashing is minimal.

During the final drying and toasting (curing) stage of malt kilning, some enzymes survive because they are intrinsically stable and can tolerate heat in a dry environment. In this stage also, the characteristic color and flavor of malt is generated primarily through a complex interaction

Flavor Descriptors for Malt Character

Note	*Thesaurus*
Cereal	Cookie, biscuit, hay, muesli, pastry
Sweet	Honey
Burnt	Toast, roast
Nutty (green)	Bean sprout, cauliflower, grassy, green pea, seaweed
Nutty (roast)	Chestnut, peanut, walnut, Brazil nut
Sulfury	Cooked vegetable, dimethylsulphide (DMS)
Harsh	Acidic, sour, sharp
Toffee	Vanilla
Caramel	Cream soda
Coffee	Espresso
Chocolate	Dark chocolate
Treacle	Treacle toffee
Smoky	Bonfire, wood fire, peaty
Phenolic	Spicy, medicinal, herbal
Fruity	Fruit jam. banana, citrus, fruitcake
Bitter	Quinine
Astringent	Mouth puckering
Other	Cardboard, earthy, damp paper
Linger	Duration/intensity of after taste

of components called the Maillard reaction or nonenzymic browning (see Chapter 3).

It is, of course, possible to heat the germinated grain without drying it. This is called "stewing." It might happen accidentally in a kiln in which airflow or temperature are too low to carry the load of moisture available and the water condenses in the grain bed or on the inside surface of the kiln (drip-back). Stewing can be easily done in a drum drier where it can be used advantageously to create special malts. Green malt, or rewetted pale malt, can be "stewed" to create colored and flavored malts including crystal malts (those malts with shiny "crystalline" endosperms) because these wet-heating conditions (1) promote mashing within each endosperm and, when the grain is later dried, (2) the Maillard reaction creates intense colors and flavors characteristic of the drying temperature used. Such malts contain no enzymes and depend on the enzymes of the pale malt, blended with them in the mash, for adequate extraction.

Well-modified malt is easy to mill and extract (see Chapter 9) in mashing. Most mashes these days are temperature programmed to produce a series of temperature stands or holds to favor the action of various enzymes,

Relevant Components of a Malt Specification

Brewers have a tendency to pile more and more line items into their specifications for malt, even though some may be mutually conflicting. For example, the best way to get low levels of DMS precursor is to restrict embryo growth and/or use a robust kilning regime. The former may lead to unacceptable undermodification and the latter to excessively high color.

Appropriate line items that should be specified in a malt are,

(1) Variety
(2) Modification and homogeneity
(3) Total protein (T)al soluble (S) protein ratio (S/T)
(4) Total β-glucan
(5) Hot-water extract (HWE)
(6) Filtration performance
(7) Free amino nitrogen
(8) Sugar spectrum

(9) Viscosity
(10) Color
(11) Absence of Flocculation factor
(12) Nitrosamines
(13) Deoxynivalenol
(14) S-methyl-methionine (SMM) (lager malt)
(15) Absence of taints
(16) Storage time

Typical values as shown in the table:

Parameter	Lager	Ale
Moisture (% max)	4.5	3.5
Friability (%)	> 80	> 85
Homogeneity (%)	> 95	> 95
Viscosity (cP)	< 1.6	< 1.55
ß glucan (ppm)	< 200	< 150
Total Protein (% max)	11	10
Soluble Nitrogen Ratio or Kolbach Index	38–44	40–45
Hot water extract (fine grind; % min)	80	82
Fine: Coarse grind difference in Extract (%)	3–4	1–3
Color (°EBC)	3–4	5–7
Diastatic Power (°ASBC)	> 150	> 100
Nitrosodimethylamine (ppm)	< 0.1	< 0.1
Deoxynivalenol (ppm)	< 0.1	< 0.1
S-methyl Methionine (ppm)	5	< 1

particularly α- and β-amylase; this produces wort of the desired extract yield and correct composition of fermentable and unfermentable sugars. Raising the temperature of the mash to the "mash-off" temperature forces the last of the malt starch into solution to be degraded by the last of the α-amylase present, stabilizes the wort properties and substantially reduces the viscosity of the mash before it enters the lauter.

Though all the wort produced in the mash could flow through the grain bed, which would clarify the wort, this is not strictly necessary; usually wort above the grain bed can be drawn off separately before sparging begins. This should considerably shorten wort separation, with advantages, (1) because lautering is usually the rate-limiting step in brewhouse processes and fast lauter turn-around time is highly valued and (2) brewers seek to minimize extraction of spent grain by minimizing contact of hot sparge water and grain.

Goals for Genetic Modification of Barley

To date, the malting and brewing industries have shown innate resistance to the prospect of using genetically modified raw materials. However, there would be a number of worthy goals for the modification of malting barley if attitudes change:

(1) Disease resistance
(2) Improved yield
(3) Reduced cell wall polysaccharides
(4) High cell wall polysaccharide degradation potential
(5) Increased foam potential
(6) Reduced haze potential
(7) Dormancy and vigor control
(8) Modifiability
(9) Low protein accumulation
(10) Increased flavor stability potential
(11) Flavor control

ADJUNCTS

Adjuncts are a significant source of brewers' extract that are not malted materials. They contribute little else but carbohydrate. They are used primarily to make beers less satiating and more crisp, lighter in color and more stable by diluting the contribution of the malt to wort. Adjuncts are primarily grains such as corn (maize) or rice or syrup extracts made from them. Some such as yellow corn grits (YCG) or rice must be boiled in the cereal cooker to gelatinize the starch before it enters the mash; the heat of the boiled adjunct mash is used to bring up the temperature of a the main malt mash at a specified rate to the required main mash temperature and fine adjustment is made with steam. Other adjuncts can be added directly to the mash because their treatment, such as rolling or micronizing or extruding, has gelatinized the starch present so that it is ready for amylolysis in the mash. Syrups, made by hydrolysis of starch from cereals especially corn, can be added directly to the brew kettle because they require no further enzyme action; such adjuncts are highly valued because they can be used to raise the gravity of wort above normal sales gravity and can also raise the level of fermentable sugar in wort above that attainable by the action of malt enzymes alone. Syrups that contain maltose can be used if desirable; this preserves the normal carbohydrate spectrum of wort. This use of syrups in the kettle, allied to the application of amylo-glucosidase during fermentation, permits the production of the immensely popular low calorie/low carbohydrate beers.

Malts and Adjuncts

Product	Details	Purpose/ comments
Pilsner malt	Well-modified malt, gentle kilning regime not rising above ca. 85°C	Mainstream malt for pale lager beers
Vienna malt	Similar to Pilsner malt but higher N, more modification, final kilning temperature ca. 90°C	Mainstream malt for darker lagers
Munich malts	From higher protein barleys (e.g., 1.85% N), prolonged germination, low temperature (e.g., 35°C) onset to kilning to allow stewing (ongoing modification), then rising temperature regime to curing at over 100°C	For darker lager beers
Pale malt	Relatively low N (e.g., < 1.65% N), well-modified, kilning starting at ca. 60°C and rising to a final curing temperature ca. 105°C	Mainstream malt for pale ales
Chit malt	Very short germination time and lightly kilned	Permissible as adjuncts in countries such as Germany with restrictions such as Reinheitsgebot
Green and lightly kilned malts	No or restricted kilning after substantial germination	Alternate to exogenous enzymes
Diastatic malts	High N barley (especially six-row), steeped to high moisture content, long cool germination, gibberellic acid if permitted, very light kilning	High enzyme potential for use in mashing with high levels of adjuncts

Product	Details	Purpose/ comments
Smoked malts	Kilning over peat	For beers with smoky character, e.g., rauchbier
Wheat malts	Germinated wheat, usually somewhat undermodified, lightly kilned (e.g., < 40°C)	For wheat beers
Rye malts		For specialty beers
Oat malts		For specialty beers, including stouts
Sorghum malts	Steeps may incorporate antimicrobials such as caustic; warm germination (25°C)	For sorghum beers; Malted millet may also be used as a richer source of enzymes
Cara Pils (a caramel malt)	The surface moisture is dried off at 50°C before stewing over 40 minutes with the temperature increased to 100°C, followed by curing at 100 to 120°C for less than 1 hour	To afford color and malty and sweet characters to lighter beers
Amber malt	Pale malt is heated in an increasing temperature regime over the range of 49 to 170°C	To afford bread crust, nutty characters to beer and color
Crystal malt	As for Cara Pils, but first curing is at 135°C for less than 2 hours	To afford toffee, caramel characters to beer and color
Chocolate malt	Lager malt is roasted, by taking temperature from 75 to 150°C over 1 hour, before allowing temperature to rise to 220°C	To afford chocolate, roast, coffee, burnt, bitter characters to beer, plus color
Black malt	Similar to chocolate malt, but the roasting is even more intense	To afford harsh, astringent, roast, burnt notes to beer plus color
Roasted barley		To afford sharp, dry, burnt and acidic notes to darker beers. (n.b. drier than roasted malts)
Raw barley		Added in mashing as a cheaper source of extract
Torrefied barley	Barley heated to 220–260°C	Easier to mill than raw barley and starch is pregelatinized
Flaked barley	Grain rolled immediately after torrefaction	Does not need to be milled
Raw wheat		Adjunct for wheat-based beers
Torrefied wheat	Wheat heated to 220–260°C	Easier to mill than raw wheat and starch is pregelatinized; wheat-based adjuncts may be used for barley malt beers to enhance foam
Flaked wheat	Grain rolled immediately after torrefaction	Does not need to be milled

Product	Details	Purpose/ comments
Wheat flour	Fine fraction produced in milling of wheat	Mash tun adjunct not requiring milling
Corn grits	Produced by milling of degermed corn (maize)	Add to cereal cooker for gelatinization; sometimes for economic reasons, or for production of lighter flavored and colored beers
Corn flakes	From torrefaction and rolling of corn	Does not need to be milled or cooked in brew house
Rice grits	Produced by milling of degermed rice	Add to cereal cooker for gelatinization; for production of lighter flavored and lightly colored beers
Rice flakes	From torrefaction and rolling of rice	Does not need to be milled or cooked in brew house
Cane sugar	Refined from sugar cane	Sucrose—for addition to kettle as wort extender or beer as priming agent
Invert sugar	Cane sugar after hydrolysis to fructose and glucose	For addition to kettle as wort extender or beer as priming agent
Corn sugars	Produced from the hydrolysis of corn starch by acid and/or enzymes	Range of products for addition to kettle depending on extent of hydrolysis. At one extreme is high dextrose sugar (approaches 100% glucose) and at the other extreme is high dextrin syrup. Latter for body—very low fermentability. Former for high fermentability (e.g., in production of light beers). Most widely used is high maltose syrup—sugar spectrum reminiscent of that from conventionally mashed malt

Some brewers use non-grist-based sources of color, viz. caramels. This is banned in countries such as Germany under the terms of the Reinheitsgebot, so there they may use coloring beers, e.g., farbebier—made from extracts of roast malt that are briefly fermented and charcoal-filtered to remove burnt character. A range of extracts of roasted malts is available in which the color and flavor components have been fractionated and therefore able to be used to introduce color without malt flavors and vice versa.

|9|

Modification

Modification is what happens to barley when it is converted to malt; it is the sum total of physical (e.g., malt friability), chemical (e.g., content of FAN or sugars) and biochemical (e.g., amount and kinds of enzymes) changes that take place during malting. Modification affects processes throughout the manufacture of beer. In this chapter, we focus only on physical and chemical aspects; Chapter 10 addresses the enzymes of malt and their role in brewing.

Modification is measured traditionally by four methods:

(1) *Coarse/fine difference* of HWE: a malt sample is milled to a standardized coarse or fine grist using different mill settings. HWE of laboratory wort is measured in the usual way and expressed as a percentage value; the value for coarse-grind extract is subtracted from the value for fine-grind extract to give the c/f difference. A c/f difference value of 2% is an acceptable number; higher values indicate less well-modified malt and lower values indicate higher modification. The method works primarily because in poorly modified malt more barley cell walls remain; thus, penetration of hot water into the coarse particles is impeded more in poorly modified malt than in well-modified malt, resulting in a lower extract value.

(2) *Soluble nitrogen ratio* (SNR), also called S-over-T, soluble-over-total nitrogen (S/T or SN/TN) or the *Kolbach index* (KI, when results are

based on an EBC Congress wort), is the amount of nitrogen in wort measured by the Kjeldahl or Dumas method (see Chapter 1), divided by the nitrogen content of the malt from which the wort was made. These ratios are expressed as a percentage. Numbers for S/T derived from boiled wort, known as the PSN or permanently soluble nitrogen, are slightly lower. For known barleys and malts of normal nitrogen content (which affects this index), values of about 40% to 45% are satisfactory; lower values imply less modification and below 35% implies poor modification. The method works because during malting, barley protein is broken down to more soluble forms of nitrogen that accumulate in the endosperm; more breakdown happens in more modified malt.

(3) *Cold water extract* is the amount of extract recovered from finely milled malt when extracted in a standard manner with ice-water, or with a dilute solution of ammonia, to prevent enzyme action during extraction; it measures the amount of readily soluble and generally low molecular weight materials in malt (mostly sugars) without the amylolytic action of the HWE method. A value of about 18% to 20% is normal and lower values imply less well modified malt; this method is losing popularity. The method works because during malting large molecules that are relatively insoluble are broken down to smaller ones that are more soluble; more breakdown happens in more modified malt.

(4) *Malt friability* can be determined directly in a friabilimeter. This device comprises a small rotating drum that is also a sieve (i.e., the drum has holes); a roller grinds the malt against the rolling sieve. If about 80% or more of the malt sample passes through the sieve, the malt is well modified, but 75% or less is poorly modified. If 1% or more of the undermodified portion of the malt fails to pass through a 2.2 mm sieve, the malt is poorly modified and has substantial "heterogeneity." The method works because undermodified portions of the grain cannot be reduced by the roller to particles that are small enough to escape the friabilimeter drum.

The following are also useful indicators of modification but are not general or standard methods: total β-glucans, β-glucans visualized with calcufluor or methylene blue stain on longitudinal sections of the grain, laboratory wort viscosity, acrospire (shoot) length (over 80% should be 3/4 to 1/1 the length of the grain in well-modified malt) and possibly the number of sinker kernels (the fewer the better).

In such analyses, the value is measured on a sample of malt that is usually 50 grams, i.e., about 1500 individual kernels. Therefore reliable and useful numbers can be gained only if the sample is taken randomly, is of sufficient initial size (or frequency) and is properly mixed and split so that the analyzed

Assessment of Cell Wall Modification in Malt

Assessment of cell wall modification is usually performed by measuring the amount of residual β-glucan in the malt. One of the ways of doing this is to sand 100 kernels that have been embedded in plastic, taking them down to about half their width and fully exposing the starchy endosperm. The plates are then flooded with Calcofluor, an agent that will bind to β-glucan selectively and register fluorescence. If there is no glucan, there is no fluorescence. The plates are viewed under ultraviolet light and the proportion that appears dark gives the percentage modified. By assessing the degree of residual fluorescence in each kernel, an estimate can be obtained of the extent to which homogeneous modification has been achieved.

One critical factor that is not always controlled is the orientation of the grain and the severity with which it is ground-down: a single grain displays different apparent extents of modification depending on how it is analyzed. If sanding is down to level A (see Figure 9.1), a large part of the endosperm will appear to be well modified. If however the sanding is done more severely, then if we are at line B, the conclusion will be that we will have a grossly undermodified grain. The same argument explains how the orientation of the grain makes a difference. If the grain is embedded with the ventral side up in the plastic, a shallow grinding will suggest undermodification, whereas a heavier shaving will register overmodification.

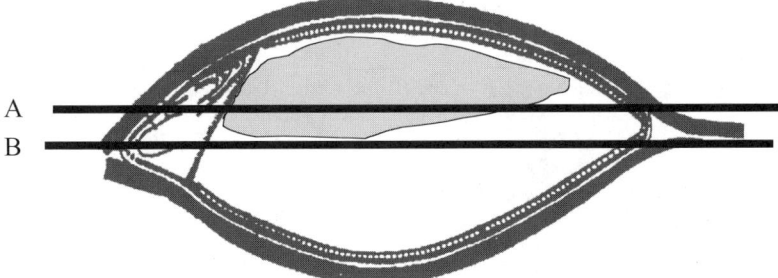

Figure 9.1. The significance of different depths of sanding in the measurement of modification of the starchy endosperm.

sample is representative of the whole mass. This is a particular challenge with solid samples such as malt or hops (see Chapter14).

Modification changes the physical qualities of the germinating grain primarily by dissolving the cell walls of the endosperm. The necessary enzymes arise in the aleurone layer. The active agent is a battery of β-glucanase and, likely pentosanase enzymes, including endo- and exoenzymes with the potential to degrade β-glucans to glucose and pentosans to arabinose and xylose. Also involved are carboxypeptidase enzymes, probably acting as

Cell Wall Degrading Enzymes from Malted Barley

Enzyme	Mode of action	Products
Feruloyl esterase	Releases ferulic acid from ester-attachment to pentosan	Ferulic acid, deesterified pentosan
Xyloacetyl esterase	Releases acetic acid from ester-attachment to pentosan	Acetic acid, deesterified pentosan
Carboxypeptidase	Hydrolyses ester linkages that have not yet been conclusively identified—perhaps between polysaccharide and peptide	Soluble β-glucan
Endo-β1-3, 1-4-glucanase	Hydrolyses β1-4 linkages adjacent to β1-3 bonds on the nonreducing side	Oligosaccharides containing three or four glucosyl residues
Endo-β1-4-glucanase	Hydrolyses β1-4 linkages located in the cellulosic regions of glucan that comprise 10 or more sequential β1-4 linked glucosyls	Lower molecular weight glucans
Exo-β-glucanases	Hydrolyze β1-4 and β1-3 bonds starting from the nonreducing end of glucans	Glucose
β-Glucosidases	Hydrolyze β1-4 and β1-3 bonds in β-linked oligosaccharides	Glucose
Arabinofuranosidase	Hydrolyses α1-2 and α1-3 linked arabinose units linked to the xylan backbone of pentosan	Arabinose, xylan
Endo-β1-4-xylanase	Hydrolyzes the β1-4 linkages within the xylan backbone	Xylooligosaccharides
Exo-β1-4-xylanase	Hydrolyses the β1-4 linkages from the nonreducing end of xylan and xylooligosaccharides	Xylose

esterases (a common property of such enzymes). This combined enzyme action is not equal (or even) throughout the endosperm of the grain because many factors intervene, but generally cell walls survive toward the distal or distal/ventral region of the endosperm. This is sometimes called a "hard end" or "steely tip." Good malting barley comprises roughly 3.5% β-glucan and malt about 0.5%, i.e., about 85% reduction. Extensive cell wall removal during germination gives the grain the quality of friability or breakability that is measured in the friabilimeter, as discussed above. This quality is also reflected in the HWE-coarse/fine difference value because the cell walls act as barriers to water penetration into milled endosperm and hence extract recovery.

Residual barley cell walls in malt affect milling and extract recovery in mashing. The impact of mill rollers readily breaks up the well-modified portions of the endosperm and easily reduces them to grits. Thus, thoroughly well-modified malt can be milled satisfactorily in simple mills, and, even if coarsely milled, can yield satisfactory extract. Residual cell walls, however, tend to cement together less well modified portions of endosperm so that coarse grits remain, from which extract is difficult to recover. If malt is less well modified, coarse grits need to be separated and remilled; this can be done in complex mills of which a five- or six-roll (dry) mill with oscillating screens might be typical. Though unmodified portions of the grain can be made more accessible to extract recovery in this way, and milling might then be thought of as completing the process of modification, it does expose the malt to excessive extraction of β-glucans and potentially to high wort viscosity. Thus, the possibility of fine milling of malt in complex mills does not obviate the need for sufficiently well-modified malt, and indeed, fine milling of poorly modified malt (such as might occur in a hammer mill) might be detrimental to wort and beer quality and to swift processing. Dynamic viscosity is expressed in centipoise (cP) or Pascal-seconds; wort viscosity might be from 1.5 to as high as 5.0 cP, depending on specific gravity and the composition of the grist, and beer about 1.2 to 2.0 cP. On this scale, the viscosity of water is 1.0 cP. Dextrins (as well as β-glucans) significantly contribute to beer viscosity.

Endosperm cell walls contain the β-glucans of barley and malt. When in solution, the β-glucans of barley cell walls exhibit extraordinary viscosity and brewers should never underestimate this characteristic. The extent to which these structural molecules survive malting and dissolve during mashing affects practical brewing in many ways. Darcy's Law, which describes the fundamentals of filtration processes includes a large factor for viscosity. Therefore, lautering and beer filtration are particularly affected by dissolved β-glucans; while the mash can be heated (at mash-off) before lautering to lower wort viscosity, this cannot be done with beer. In quite the opposite way, beer filtration is typically done at a very low temperature (e.g., −2°C). Beer filtration is therefore particularly susceptible to the presence of

Cell Wall Model

In 2001, a model was developed to account for the structure of the cell walls of the starchy endosperm (Figure 9.2). The bulk of the wall comprises β-glucan, but this is enfolded in the pentosan component. Although there is some free accessibility to the glucan for enzymes (the model depicts this by the "holes" in the pentosan layer), the removal of the pentosan layer and its various components (xylan, arabinose, ferulic acid and acetic acid) enables the solubilization of glucan. Furthermore, glucan can be released by the hydrolysis of linkages between polysaccharides and proteins. As a consequence the cocktail of enzymes that attack these outer layers (arabinofuranosidase, xylanase, feruloyl esterase, xyloacetylesterase and carboxypeptidase) constitute "solubilase."

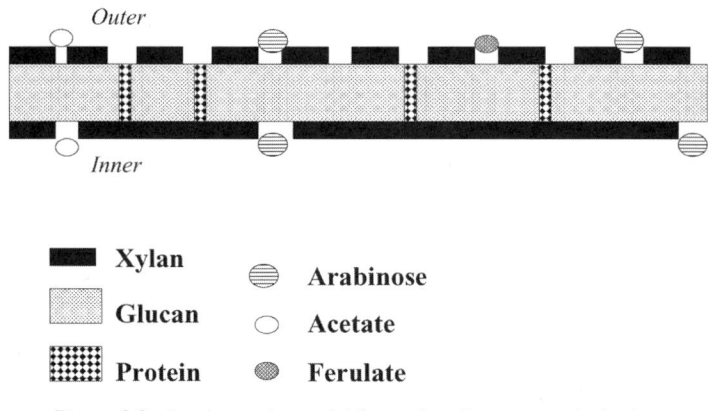

Figure 9.2. A schematic model for cell wall structure in barley.

dissolved β-glucans, and deposited β-glucans can rapidly clog fine filters, especially membrane filters. In sufficient amounts β-glucans can form hazes in beer and even deposits in the form of gels. The viscosity of β-glucans might have some positive effect on the quality of beer foam, e.g., its wetness and rate of drainage, but such an amount would negatively impact filtration (see Chapter 4). β-Glucans and pentosans of beer are classed as soluble fiber, which is regarded as a beneficial food component. Removal of β-glucans contributes to the drinkability (refreshing or nonsatiating quality) of beer.

Darcy's Law

Darcy's law brings together the factors that have an impact on the rate at which liquid will flow through a particulate system:

Rate of liquid flow = pressure × bed permeability ×
filtration area/bed depth × wort viscosity

Wort is collected more quickly in a shallow system with a large surface area and where there is a possibility to force liquid through with pressure (provided this does not compress the bed and reduce permeability). Low-viscosity liquids flow more readily.

The permeability factor relates to the characteristics of the solids in the system. It is substantially dependent on particle size; liquid flows readily around larger particles but tends to be held up if it is flowing around much smaller particles.

Beer as a Source of Nutrients

Barley β-glucans and pentosans are generally regarded by brewers as problematic and so the goal is their extensive removal during malting and brewing. However, complete conversion of glucan to glucose does not occur and significant quantities of low-molecular-weight β-linked oligosaccharides are generated. Furthermore, there are even high levels of partially degraded arabinoxylans surviving into beer.

These molecules are neither metabolizable by brewing yeast (hence they enter into beer) nor are they metabolizable by the human. As such, they probably survive into the large intestine and thus represent a part of the fiber component (according to the above definition) or may be considered as prebiotics. The latter are lower molecular weight substances that serve as substrates for beneficial organisms in the colon.

So, some beers may yield significant amounts of fiber and prebiotics to the diet. Other useful substances that are found in beer, sometimes in

> very significant quantities, are B-vitamins (except thiamine), inorganic ions such as selenium, silicate, calcium and phosphate, and antioxidants including polyphenols and ferulic acid.
>
> However, perhaps the most significant component is ethanol. Apart from being a substantive calorific source, it is almost certainly the active ingredient involved in reducing the risk of atherosclerosis in the body.

The time to deal with β-glucans is during malting (see Chapter 8). If this is not done, i.e., the malt is not well-modified, the brewer has only four possible options and none of them is particularly attractive: (1) Assure that the residual malt β-glucanases have the opportunity to act, e.g., during the low-temperature stages of a temperature-programmed mash. Unfortunately, undegraded β-glucans might dissolve after the mash temperature has risen too high for β-glucanase survival; (2) Add a commercial β-glucanase preparation (preferably a relatively heat-stable one) to the mash. (3) Mill and mash in such a way as to minimize β-glucan extraction (though this would tend to sacrifice brewhouse yield also). Or, (4) possibly, apply appropriate technology, e.g., use thin-bed mash filters with inflatable membranes and substitute centrifuges for filters. In breweries where milled barley is used as an adjunct, only options (2) and/or (4) are effective. If the brewer's concern is only with beer filtration, β-glucanase can be added to beer.

Malt modification affects yeast nutrition and performance during fermentation, and beer flavor and stability. Amino acids are in short supply in barley but ample for brewers' needs in malt. Though it is easy, these days, to measure the spectrum of individual amino acid in malt, this is rarely done, although amino acids might be quite variable among malt samples. The amino acid potential is, instead, measured on wort and reported as FAN, although peptides, polypeptides and even proteins can give a positive value in this test. The method uses ninhydrin (1,2,3-tri-keto-hydrindine hydrate) to yield a blue color in solution (the reaction, incidentally is a special case of the Strecker degradation notable in color-forming reactions between amino acids and sugar derivatives; see Chapter 3). Values for FAN in malt or wort of 150 to 250 mg/100g or mg/l, respectively, are appropriate.

No significant fermentation takes place in the absence of yeast growth, and growth requires amino acids because the nitrogen of these molecules is necessary for forming protein of new yeast mass. The flavor compounds that yeast contributes to beer are the waste products of metabolism and more metabolism; i.e., generally speaking, more growth, promotes more flavor compounds. Although yeast growth does not respond in a linear way to the FAN of wort, amino acids affect the amount of growth and should therefore be consistent. Specific amino acids also affect flavor directly: their deaminated and decarboxylated carbon skeletons contribute to the higher alcohol spectrum of beer (see Chapter 11). Amino acids are also buffers, i.e., resist change in pH (see Chapter 2) and help establish the general pH environment of brewing. Particularly, aspartic acid and glutamic acid contain an ionizable side chain with a pK value of about 4.0, and so these amino acids, and the peptides that contain them, are the most effective buffers in brewhouse processes, wort and especially beer. They affect enzyme action in mashing and beer pH, for example, and so (with CO_2) beer mouthfeel. We have previously noted the participation of amino acids in color formation during kilning and wort boiling (see Chapter 3). Because more fully modified malt generally contains higher FAN and CWE, they more readily form color compounds in these processes. Amino acids and small peptides that survive into finished beer have the potential to exacerbate the growth of unwanted microbes in beer, especially the lactic acid bacteria, some of which are quite demanding on amino acids for nutrition. Finally, it is plain that in well-modified malt, FAN represents the breakdown of barley proteins; in general, this can be assumed to favor physical (chill-haze) stability and disfavor foam stability (see Chapters 4 and 5).

The small starch granules of barley endosperm are absent in well-modified malt endosperm because they are easily and completely degraded to sugars by amylases during malting. However, small starch granules that survive malting tend also to survive mashing because they are less easily solubilized than large starch granules under mash conditions. If present they can react with some proteins, polysaccharides and lipids to retard wort separation during lautering.

Starch-Degrading Enzymes from Malted Barley

Enzyme	Mode of action	Products
α-Amylase	Hydrolyzes α1-4 linkages inside amylose and amylopectin	Oligosaccharides
β-Amylase	Hydrolyzes alternate α1-4 linkages starting from the nonreducing ends of amylose and amylopectin molecules, stopping when encountering α1-6 side chains	Maltose, "β-limit dextrins"
Limit dextrinase	Hydrolyzes α1-6 linkages inside amylopectin	Linear dextrins
α-Glucosidase	Hydrolyzes α1-4 linkages of oligosaccharides	Glucose

Malt also provides vitamins and minerals to worts (and ultimately beers) that are important for yeast nutrition. Potassium and phosphate are quantitatively dominant ions, the latter doubtless derived mostly from phytate by action of the phosphatase enzyme(s), "phytase," during malting. Phosphate and phytate react with calcium of brewing water during brewhouse processes to release H^+ ions (acidification), and establish part of the normal pH environment for wort production (see Chapters 2 and 7). Phosphate has many uses in intermediary metabolism (e.g., as ATP in glycolysis) and in membrane structures (e.g., as phospholipids). Much smaller, but essential, amounts of other elements, especially zinc and magnesium, also arise from malt. Malt contains ample zinc for brewing purposes but it is poorly extracted into wort; supplementing worts with zinc is therefore quite common. Dephosphorylation of phyate also releases *myo*-inositol (a B-vitamin) that also plays an important role in membrane structures of the yeast cell. Inositol and nicotinic acid are the dominant vitamins contributed from (as might be expected) the living tissues of barley, along with nutritionally important amounts of the other B-vitamins; thiamine, for example, has a role to play in pyruvate decarboxylase and pyridoxine in amino acid transaminase reactions. Vitamins increase somewhat with modification during barley germination, which again links good modification to the nutritional quality of wort for yeast growth and fermentation. It is axiomatic, however, that a properly made malt wort, without excessive dilution with adjunct, will be nutritionally sufficient in sugars, amino acids, vitamins and minerals for yeast growth.

Although lipids are extensively metabolized during modification, barley and malt contain roughly the same amount of lipid (about 3.5% and 3.0%, respectively) and modification cannot therefore be said to greatly influence the effects of lipid in beer quality. Though some small amount of lipid might be beneficial to yeast nutrition, especially unsaturated fatty acids and some sterols for the formation of sound membrane structures under anaerobic conditions, lipids, in general, can have antifoaming properties and if oxidized contribute to stale flavors. Fortunately, most (some 80%–90%) of malt lipid remains in the spent grain.

Finally, the precursor of DMS, called SMM, is formed during malt modification because it is important in metabolism for the transfer of 1-C (methyl groups) in the developing embryo. All else being equal, more modification favors SMM in malts, and so undermodification is one method for control of SMM. SMM is broken down to DMS and dimethyl sulfoxide (DMSO) in hot processes, such as kilning, especially at high curing temperature, and wort boiling, and is removed by normal evaporation that occurs in those processes. However, holding hot wort without evaporation, favors the conversion of SMM to DMS and DMSO and their survival into beer. Manipulation of hot processes and concomitant evaporation are therefore the routine practical ways of SMM/DMS control in breweries.

Nitrosamines

A study in Canada in the 1970s revealed that malts contained nitrosamines, potential carcinogens. Notable amongst them is nitrosodimethylamine, which is produced during kilning by the reactions between oxides of nitrogen (NO_x) in the air and a precursor substance, hordenine, found in the embryo. Therefore, procedures that minimize embryo development (e.g., use of bromate or ammonium persulfate) are helpful in reducing nitrosamine levels. More importantly, elimination of NO_x is important. This is achieved by using indirect heating during kilning—i.e., instead of passing hot air directly through the malt, heat is transferred through some sort of heat exchanger that avoids contact of malt with the air. An alternative is to acidify the malt because the reaction does not occur under acidic conditions. This has been achieved by burning sulfur on the kiln, and dissolving the sulfur dioxide and sulfur trioxide dissolve in moisture to give acid. However, the kiln is quickly corroded.

Control of Dimethyl Sulfide (DMS)

All of the DMS in a lager ultimately originates from a precursor, S-methylmethionine (SMM), which is produced in the embryo of barley during germination. As it is based on an amino acid, there are higher levels in barleys containing more protein. It is also developed in higher amounts in six-row barley. Factors that promote modification increase levels of SMM, whereas anything that suppresses embryo growth—e.g., potassium bromate—lowers levels of SMM. As barley matures after storage, it increases in its ability to develop SMM during germination.

SMM is heat-sensitive and is broken down rapidly whenever the temperature gets above about 80°C. Accordingly, SMM levels are lower in the more intensely kilned ale malts, which is a major reason why DMS is more associated with lagers. SMM is extracted into wort during mashing and is further degraded during boiling and in the hot wort stand. The half-life of SMM at 100°C is 38 minutes, and this value doubles for every 6°C decrease in temperature. Thus there is no significant splitting of SMM during infusion mashes, but will be during the boiling stages of decoction mashing. If the kettle boil is vigorous, most of the SMM is converted to DMS and this is volatilized. In the whirlpool, the conditions are much less turbulent than in the kettle; any SMM surviving the boil will be broken down to DMS but the latter tends to stay in the wort. Brewers seeking to retain some DMS in their beer will specify a finite level of SMM in their malt and will manipulate the boil and whirlpool stages in order to deliver a certain level of DMS into the pitching wort. During fermentation much DMS is driven off with carbon dioxide, and so the level of DMS required in the wort will be somewhat higher than that specified for the beer. However, there is also some production of DMS during fermentation. This is due to the reduction of DMSO to DMS by yeast. Some of the SMM is converted during kilning into DMSO, especially in the higher temperatures during curing. DMSO is not heat labile but is water soluble. It enters into wort at quite high levels; yeast contains an enzyme called sulfoxide reductase that reduces DMSO to DMS. The preferred substrate for this enzyme is methionine sulfoxide (MetSO) because the *in vivo* role for the enzyme is to keep methionine in an unoxidized state by turning MetSO immediately back to methionine. MetSO is present in ale and lager malts. It, too, is produced in the curing stage of kilning. Both MetSO and DMSO are present at higher levels in ale malts than in lager malts because of the higher kilning temperatures as compared to lager malts. Because MetSO delays the reduction of DMSO, this means that there is less tendency for DMS production by yeast in ale fermentations than in lager fermentations. Yeast strain is very important, some strains producing much more DMS than do others. Other significant factors are the level of assimilable nitrogen in wort—it is only when the nitrogen level is depleted that DMSO reduction occurs to a significant extent—and temperature, with much more DMSO being reduced to DMS at, say, at 8°C than at 16°C. Finally, certain bacteria are far more adept at reducing DMSO, than is yeast. Whereas yeast might reduce only 5% to 10% of DMSO to DMS, certain gram-negative bacteria such as *Obesumbacterium proteus* can completely reduce DMSO to DMS very rapidly; this is why such wort spoilers tend to generate strong DMS aromas.

|10|

Enzymes

Breweries have increasingly become the bailiwick of engineers because of their immense size, complexity and widespread electronic monitoring and control. Nevertheless, making beer remains crucially a biological process. Enzymes, which are catalysts made only by living tissues, are the exemplar of biology; the three essential transformations of beer making, i.e., modification of barley to make malt, extraction of malt to make fermentable wort and fermentation of wort to make beer, are all catalyzed by enzymes. The brewer's main task is to assure consistent control of these processes by managing enzyme action.

Enzymes are biological catalysts. All catalysts accelerate thermodynamically possible chemical reactions. Enzymes however have two advantages over chemical catalysts. First, enzymes allow reactions to proceed under modest (to us) conditions of temperature, pH, concentration and pressure; this makes possible the complex series of chemical reactions that we recognize as life. Second, enzymes have specificity of action; that is, enzymes conduct only one reaction, or only one kind of reaction, and direct it in a stereo-specific way. Any enzyme reaction comprises three elements: the enzyme itself, the substrate or compound being acted upon, and the environment or conditions under which the reaction takes place; each has a central bearing on the rate and extent of chemical change achieved.

Assay of Enzymes

Enzymes are customarily measured on the basis of the rate of the reactions that they catalyze. There is (within normal circumstances) a linear relationship between the rate of a reaction and the amount of enzyme present. Therefore, providing other parameters that impact the rate of a reaction are kept constant (substrate concentration, pH, temperature, presence of inhibitors and activators), then the rate of the reaction forms the basis of reliable quantification.

The reaction may be monitored by following either the decrease in level of a substrate or increase in the level of a product. In either instance, enzymologists usually measure the "initial rate" of a reaction because, sooner or later depending on the enzyme concerned, there is a falling off of the rate at which the reaction occurs.

Assay is usually conducted in isolation from the system in which an enzyme normally operates. For example, β-glucanase can be measured using an artificial substrate that comprises β-glucan attached to a dyestuff. When the enzyme attacks this substrate it releases the dye, which can be measured through its absorbance. Within the barley grain during germination or in a mash, however, the β-glucanase operates alongside a diversity of other enzymes. For example, the complex of enzymes known as solubilase releases glucan from attachment to cell walls and the glucanase then attacks these. Further, there are exoglucanases and glucosidases that act on the products made by the glucanase. Thus there is a very dynamic scenario. The same applies to the enzymes dealing with proteins, starch and lipids. Thus it is not easy to extrapolate from enzyme assays to the events in situ. However assays serve a useful purpose in quantifying enzymes for purposes such as comparing malts and commercial enzymes.

However the most useful assays are the ones in which the enzyme is monitored in scenarios as close as possible to the situation in which it is expected to act in the brewery. This applies especially to commercial (exogenous) enzymes). For example, an enzyme may have very high activity in an assay performed with an artificial substrate under "optimum" conditions in a test-tube. However, the important thing to know is whether the enzymes will function to valuable effect in the brewery. To take our example of a β-glucanase again, then the most effective enzyme is not necessarily the one that works well with a dyed substrate in vitro, but rather one that will actually lessen the viscosity in a mash, under the conditions of water: grist ratio, temperature and times that prevails. Thus the best "living assays" are the ones that recreate commercial practice: a useful assay for a β-glucanase, then, involved small-scale mashes, with addition to separate beakers of different levels of the enzyme and a monitoring of the rate of filtration of the mashes, and measurement of viscosity and the level of Extract in the resultant worts.

A selection of assays used for monitoring enzymes of relevance to the maltster and brewer:

Enzyme	Assays
β-Glucanase	1. Change in absorbance resulting from hydrolysis of dyed substrate 2. Clearing of color from glucan dyed with Congo Red. The dyed substrate can be immobilized in agar plates. Holes are punched in the agar and the enzyme put into the wells. The enzyme diffuses into the agar and zones of clearing appear. The bigger the diameter, the more enzyme 3. Decrease in viscosity of standardized β-glucan solutions
α-Amylase	The substrate has been treated with β-amylase and also an excess quantity of that enzyme is added to the assay. The "β-limit dextrin" substrate is mixed with the malt extract and the rate of loss of iodine staining potential is assessed. The more quickly the color is lost, the more α-amylase is present. Activity is cited as "dextrinizing units"
α-Amylase + β-amylase	Together they are assessed as *diastatic power*. An extract of malt is incubated with a standard starch solution. The reducing sugars produced are measured.
Proteinase	There are numerous assays for proteinase but they mostly involve artificial substrates such as dye-linked substrates that are not native to brewing materials e.g. hemoglobin. The best substrates would be the ones found in grain (e.g., hordein) but these are not readily available commercially. Furthermore there are numerous proteinases found in malt and they differ in their substrate specificity.
Carboxypeptidase	An artificial dyed dipeptide is used as substrates: when the enzyme hydrolyzes it there is a release of dye that can be monitored in a spectrophotometer
Lipoxygenase	When the enzymes oxidized linoleic acid, a change in ultraviolet absorbance can be monitored

Though enzymes accelerate reactions by a factor of millions, or often more, they accelerate the forward and reverse reactions to the same extent so that although the equilibrium of the chemical reaction is reached very quickly, the equilibrium of the reaction is not altered, i.e., the equilibrium constant of a reaction is the same enzymatically and nonenzymically. However, some enzymes make possible changes in chemical structure that proceed with glacial slowness nonenzymically and catalyze barely perceptible reverse reactions; in this sense enzymes might be said to cause specific chemical changes. Hydrolysis of starch by amylases is a good example of this.

For all practical brewing purposes enzymes are proteins, that is comprising a twisted and folded backbone chain of amino acids constructed in specific order (see Chapter 1). These protein structures often include additional components that are essential to their enzyme activity and can be more or less easily removed from proteins. These are commonly metal ions (e.g., Mg^{2+} of kinase enzymes, essential in glycolysis, or Zn^{2+} of aldolase and alcohol dehydrogenase) or vitamins (e.g., thiamine of pyruvate decarboxylase or pyridoxine of transaminases).

Enzyme proteins, as all proteins do, depend on their structure to define their function. The protein character of enzyme molecules allows them to bind to the substrate being transformed; this is a necessary feature of enzyme action. Binding between substrate and enzyme takes place at the active site(s) of the enzyme; this site(s) has an unique structure that can be thought of as a three-dimensional crevice or cup in, or on, the enzyme molecule. The site(s) arises from the complex twisting and folding of the primary chain of amino acids. Herein lies the specificity of enzymes: the active site(s) are conformed to fit or accept only certain molecules (highly specific enzymes) or certain kinds of molecules (enzymes with group specificity) to the exclusion of all others. Binding takes place through the extraordinary ability of proteins to form hydrogen bonds. The hydrogen donor and hydrogen acceptor necessary to form a hydrogen bond can be on the protein and/or on the substrate.

pH also affects binding. The protein character of enzyme molecules also causes one of the more intriguing characteristics of enzymes, that is, their sensitivity to pH and their pH optimum. The pH environment (surrounding hydrogen ion concentration) of any protein influences the dissociation of weak acid and weak base groups of the amino acids that comprise the protein molecule; of particular importance are the ionizable groups in the amino acid side chains such as the –COOH of aspartic and glutamic acid and the base N-containing terminal groups of, e.g., lysine and arginine. Ionization of these groups determine the local electrical character of a protein, e.g., at the active site(s), as well as its overall charge. For example, if the positive and negative charges balance each other the protein is said to be at its isoelectric point. This charged character provides another mechanism for specific binding between substrate and enzyme, by electrostatic interaction or ionic bonding (opposite charges attract).

The binding of substrate to enzyme is the essential step in the conversion of substrate to product. The binding creates a new environment around the substrate that effectively lowers the amount of energy required to drive the reaction forward (the activation energy of the reaction). When bound to an enzyme, sufficient energy can be derived from the heat of the surroundings to drive forward the enzyme-catalyzed reaction, though such energy is quite insufficient to cause a reaction under nonenzymic conditions.

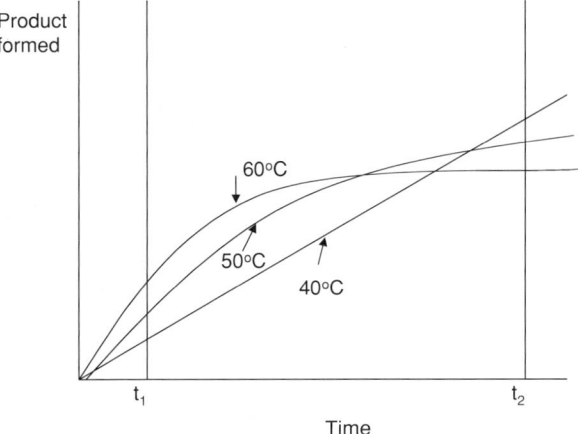

Figure 10.1. Product formation in an enzyme-catalyzed reaction at different temperatures.

The rate of enzyme action depends on several factors: the amount of enzyme, and hence number of active sites present, being one. In the presence of excess substrate the enzyme is said to be saturated and every active site starts working as fast as it can; in such case the overall rate of enzyme reaction depends on the concentration of enzyme present. As time goes by, however, the reaction slows down and eventually stops either when equilibrium is approached (the action has not stopped in this case; the forward and backward reactions are equal) or when all of the substrate is mostly converted to product. In practice, other reasons affect the rate of reaction and the point at which it stops. As the substrate is used up the active sites are inadequately supplied with substrate (they are said to be unsaturated) and so the reaction slows. In addition, the enzyme itself is inactivated by heat (Figure 10.1) or by inhibitors that accumulate (including in some cases the product of the reaction, so-called product inhibition), which also slows the reaction. The action of amylases on starch in the brewer's mash is again a good example of these concepts.

The protein character of enzyme molecules is also their Achilles heel because loss of structure (called denaturation in the case of proteins) means loss of function, i.e., loss of enzyme activity. Denaturation is generally portrayed as a partial uncoiling or unraveling of the complex three-dimensional structure of the molecule caused most commonly in malting and brewing processes by heat, such as kilning, mashing/boiling and pasteurization, but also by extremes of pH. Once denaturation happens and enzyme activity is lost in brewing processes, it cannot be recovered. For this and other reasons temperature control is a central concern in brewing processes.

Most chemical reactions are endothermic, i.e., they require heat, and enzyme-catalyzed reactions are no exception: over a modest range of

"Temperature optima"

Too often, people speak of "temperature optima" for enzymes. This is a risky concept that pertains to the way in which enzymes are assayed.

Figure 10.1 illustrates the rate of formation of products catalyzed by a hypothetical enzyme at three different temperatures. As is the case for all chemical reactions, an increase in temperature makes for a more rapid reaction. However, enzymes display varying extents of sensitivity to heat. As the temperature increases, so does the rate of inactivation of the enzymes. So at the highest temperature tested there is a more rapid inactivation of the enzyme, and a tailing off in the rate of reaction. However, at the lower temperature the enzyme survives and continues to perform the reaction, ultimately producing more product than the enzyme can when working at the highest temperature (the classic "tortoise versus hare"). Imagine that we assay the enzymes measuring how much product is made after a set period of time (Figure 10.1). If we set that period of time at t_1, then our conclusion would be that the "temperature optimum" for the enzyme would be 60°C. If we set the assay time at t_2, then our conclusion would be that the temperature optimum is 40°C. Let us say that the enzyme was a β-glucanase and that the two assay points were 5 minutes and 60 minutes. Imagine if the enzyme supplier assayed using a test tube assay at the 5-minute interval: they would tell us that the enzyme worked best at 60°C. In fact it would be killed off in the assay after 20 to 30 minutes, likely before it had completed its job.

temperature, higher temperature accelerates enzyme action. However, depending on the enzyme, some temperature will eventually be reached at which the reaction no longer accelerates but slows down due to the thermal inactivation of the enzyme. Thus, *under a given set of operating parameters*, an optimum temperature can be defined for every enzyme. The optimum temperature is not an intrinsic property of the enzyme molecule (as the pH optimum is) but varies with the conditions and objectives of the reaction; thus, in the brewer's mash, for example, the temperatures used cause profligate enzyme destruction to achieve sufficient enzyme action in a short time.

Enzymes participate in three general kinds of reactions: those that degrade the substrate to small molecules (e.g., β-glucanases, amylases and proteases), those that build up small molecules into large ones (e.g., enzymes responsible for yeast growth) and those that create the energy necessary for that synthesis (e.g., some enzymes of glycolysis). Enzymes generally act

Various Types of Enzyme Inhibition

Enzymes can be inhibited in several ways and some of these are relevant in a brewing scenario.

If there is an excessive amount of substrate present the separate molecules will mutually interfere with their access to the active site of the enzyme. This is called *substrate inhibition*.

If there is an accumulation of product at or near the active site, such that the access of fresh substrate is impeded, then it is called *product inhibition*.

If a nonsubstrate molecule has a structure and shape very similar to the substrate then it may be able to interact with the active site and "squeeze out" the substrate. This is called *competitive inhibition*.

Other nonsubstrate molecules may interact with enzymes at sites remote from the substrate, but distort the enzyme structure such that the active site is deformed and unable to work. This is called *noncompetitive inhibition*.

There are several specific inhibitors in cereals that function to block enzyme action as part of control systems within the grain. Such inhibition may either be competitive or noncompetitive. Examples of such *endogenous inhibitors* are those that block limit dextrinase and xylanase.

All of the inhibitions referred to above are reversible, in that removal of the inhibitor allows the enzyme to perform. Irreversible inhibition occurs when a molecule combines with an enzyme to block a reaction, and can either not be removed or, if it can be removed, it has irreversibly interfered with the ability of the enzyme to function. These molecules are usually referred to as *inactivators*, and the most important examples in a brewing context are heavy metal ions such as copper and perhaps high levels of polyphenols.

within living tissue, for example, in the fermenting yeast cell and in the aleurone and embryo of malt, where they are under control, by various control mechanisms; as a result, enzyme action is quite efficiently balanced with the needs of the living cell. In such case enzymes are usually organized into pathways so that the product of one enzyme becomes the substrate for another. Although these pathways are pictured in textbooks as sequences, it is not necessary for enzymes to be lined up in this way in nature, e.g., organized onto some cell substructure (though some are); Glycolysis, for example, the essential pathway of 11 enzymes that converts glucose to pyruvate, is located in the cytoplasm of the yeast cell from whence it may be extracted and glycolysis achieved, cell free, in a test tube. Substrate specificity of enzymes permits this. Enzymes can also act in nonliving environments such as in the endosperm of the barley kernel (in the presence of water during germination) and in the brewer's mash. In such case enzyme action is controlled by the conditions under which action takes place, particularly enzyme concentration, temperature, substrate availability, water and pH. This is the essential fact that puts, e.g., endosperm modification (see Chapter 9) and wort composition substantially under the control of maltsters and brewers.

Maltsters and brewers depend on only one standard measure for the enzymic quality of malt: the determination of diastatic power (DP). This measures, under exactly prescribed conditions that must be followed to the letter, the combined action of α-amylase and β-amylase of a specified malt sample. An extract of malt is diluted and reacted with a dilute solution of a specially prepared standard starch for exactly 30 minutes at exactly 20°C. The amount of reducing substances formed is then measured using iodimetry (the reducing substances reduce ferricyanide to ferrocynaide and the unconsumed ferricyanide is used to release iodine form KI solution). The iodine formed is then quanitified with a standard sodium thiosulfate solution. The net volume of thiosufate used, 50×, is the malt DP, a dimensionless number. The DP scale is linear over the practical range of malt DP.

Commercial Enzymes Used in Brewing

Enzyme	Stage added	Function
β-Glucanase	Mashing	Elimination of β-glucans, especially when using barley and oat adjuncts
Xylanase	Mashing	Elimination of pentosans, especially when using wheat adjuncts
Proteinase	Mashing	Production of free amino nitrogen, especially in high adjunct mashes
Amylases	Mashing	Production of fermentable carbohydrate in high adjunct mashes
Glucoamylase (amyloglucosidase)	Mashing or fermenter	Yielding increased fermentable carbohydrate, for production of light and low carbohydrate beers
Acetolactate decarboxylase	Fermenter	Accelerating the maturation of beer by circumventing the production of diacetyl
Papain	Stored beer	To eliminate haze-forming polypeptides
Prolyl endopeptidase	Stored beer	To eliminate haze-forming polypeptides; potential value in producing beer for celiacs
Glucose oxidase/catalase	Packaged beer	Elimination of oxygen

Genetic Adjustment

Brewers remain wary of gene technology. Of special concern to them are adjuncts that are based on genetically modified cereals that are already widely used, e.g., corn.

Technology now allows for the genetic modification of the raw materials used in brewing. The first genetically modified brewing yeast approved for commercial use was developed more than a decade ago. This yeast was modified so as to contain the gene for glucoamylase and therefore is of value for the production of light beers. However the wariness of the brewing industry has meant that this organism has never been used commercially. Other constructs of potential future interest are yeasts that express acetolactate decarboxylase. Of more interest perhaps would be yeast that are not genetically modified *per se*, but which are mutated so that they no longer produce undesirable components, or conversely overexpress desirable substance. In the first instance we might consider yeasts that produce less hydrogen sulfide, dimethyl sulfide or diacetyl. In the latter could be yeast that produce higher levels of sulfur dioxide.

Commercially successful mutants of barley have been produced, notably the low proanthocyanidin varieties that allow increased resistance to haze formation in the beers produced from them. With regard to genetic modification of barley, this is more challenging than for microorganism but has now been achieved. As yet, no commercially viable genetically modified malting barleys have been developed.

Yeast

Hopped wort is the raw material from which beer is fermented by yeast. As a result of fermentation there is a sea change in the character of the beverage from a sweet, rather bland and satiating drink (wort) to one that has delighted humankind for millennia (beer). Four changes accrue that define this transformation; broadly they are (1) the removal of sweetness by yeast action and its replacement with alcohol; (2) the formation of acids by yeast action and removal of buffers and hence the lowering of pH; (3) carbonation caused by the formation of CO_2 by yeast action and (4) formation and release of a host of metabolic waste products, each in low concentration. In this way yeast magnificently enhances and reveals the fundamental flavor of beer derived primarily from malt and hops. It is therefore hardly surprising that brewers invest heavily in their yeast, guard it carefully in storage, analyze its quality before use and recycle yeast only from those fermentations that meet specifications. Because brewers have been recycling yeast for millennia, brewing yeasts are peculiar to breweries. Although brewing yeasts can all be categorized taxonomically under *Saccharomyces cerevisiae*, brewers understand that characteristics that are crucial to them are trivial to taxonomists (e.g., good beer flavor and satisfactory mechanical action!); therefore, brewers tend to retain traditional names such as ale yeast or lager yeast, top yeast or bottom yeast, *Saccharomyces cerevisiae* or

Yeast Taxonomy

Taxonomists seem to have struggled for a number of years with the names that should be ascribed to brewing strains. Ale yeast has long been referred to as *Saccharomyces cerevisiae* and that practice remains. It is the bottom-fermenting lager yeasts that have received different names as research has developed. Successively, they have been named *S. carlsbergensis*, *S. uvarum* and *S. cerevisiae* lager-type. Now, they are strictly termed *S. pastorianus*. It is understood that *S. pastorianus* evolved from a melding of *S. cerevisiae* with *S. bayanus*, resulting in the larger and more complex genome of lager strains.

Saccharomyces carlsbergensis because these names reflect what they have experienced about yeast. Brewers also know that their own yeast strain(s) (within the general classification of brewing yeasts) allied with raw materials and the processes used in their breweries, are essential to the unique qualities of their own beers.

In brewing practice yeast grows under very restricted conditions caused primarily by the absence of oxygen (fermentation), relatively low temperature and recycling practices. The conditions used exercise a selective pressure on the population, and yeasts become adapted to certain brewing practices under which they perform satisfactorily. Fermentation results in the inefficient extraction of energy from fermentable sugar and so, relative to the large amount of sugar and other metabolites utilized, the yield of new yeast mass is quite small. This means that a good deal of material is left behind as metabolic waste products and appears in the beer as alcohol and CO_2 primarily (along with glycerol and flavor compounds). Contrast this to aerobic metabolism where much yeast mass accumulates and the end-products are essentially CO_2 and H_2O! The brewer's task is to manipulate wort qualities and the conditions of fermentation in such a way that beer of consistent flavor quality is made efficiently. Thus, controlled yeast growth (rate and amount) is the key to successful beer production. Taking a simple mass-balance approach to fermentation inputs and outputs (Figure 11.1), it is clear that additional yeast growth must subtract from formation of alcohol/CO_2 and/or flavor compounds and vice versa.

Fermentation does not take place in a simple solution of sugar because yeast cannot grow under these conditions; additional nutrients are required for its growth. Though yeast can grow well on sugar plus ammonia and a

Chapter 11

Inputs

Carbon
Nitrogen
Oxygen
Hydrogen
etc

Outputs

Carbon dioxide
Ethanol
Miscellaneous metabolic side-products

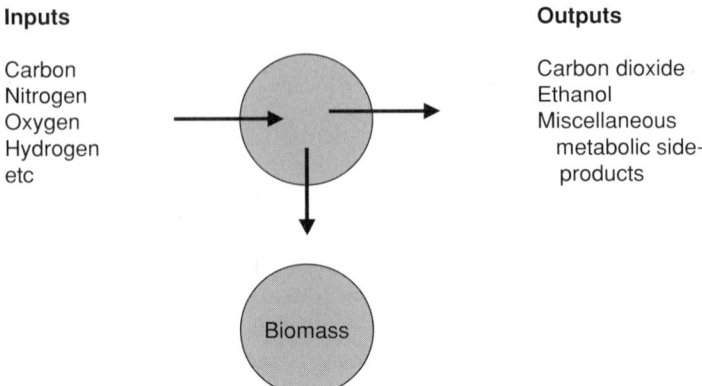

Figure 11.1. Inputs and outputs in fermentation.

few inorganic salts, growth under anaerobic conditions is much enhanced by addition of vitamins and amino acids as a source of nitrogen. Consistent and sufficient initial wort aeration is also a major determinant of yeast growth in brewery fermentations and in the performance of recycled yeast. Measurement of yeast growth is complicated in practical systems because during fermentation yeast grows but at the same time settles at the bottom of the fermenter; brewers therefore measure *yeast in suspension* at any given moment and not *total cell mass* in the system. Practical measures of "yeast growth" must therefore be interpreted carefully. Many researchers, in contrast, study fermentation in relatively small volumes in tall-tube fermenters or in ones that are stirred or agitated. Yeast growth can be assessed as packed

---

Acid Washing of Yeast

Many brewers treat yeast prior to pitching with acid (to a pH of 2.2). This is an effective way to kill bacteria, though not wild yeasts and, of course, almost always not the brewing yeast itself. It is important that food-grade acid is used (usually phosphoric), that the yeast is healthy and cold (4°C–5°C), that the slurry is well-mixed when the acid is dosed in and that the slurry should not be kept longer than 2 hours prior to directly dosing into the fermenter.

Assessing the Quantity of Yeast

It is important that the correct quantity of the correct healthy yeast is pitched into wort if the appropriate time-course for fermentation is to be pursued. Quantification may be by weight or by volume, and a customary order of magnitude is 1×10^6 cells/ml per degree Plato. Yeast numbers can be measured using a hemocytometer, which is a counting chamber loaded onto a microscope slide. Alternatively, yeast may be centrifuged in pots calibrated to relate volume to mass, remembering that there are usually other solid materials present, viz. trub.

In-line instruments are now available. One invokes capacitance measurement: living intact yeast cells operate as capacitors (they will store charge) and the extent of this is measured as a direct index of viable cell numbers (calibration being against cell number). Alternatively, the extent to which slurries scatter light is in proportion to cell number, though it is important to correct for nonyeast materials that will scatter light.

cell volume in a centrifuged sample (often treated with alkali to minimize interference from trub) or, if the cells are washed free of wort, as dry weight of cells. Light scattering of diluted samples can also be calibrated with cell mass. Growth can be measured directly by counting cells in a hemocytometer, or electronically in a Coulter particle counter or Abmeter or optically (using near-infrared (NIR) spectroscopy, for example). The gold standard is traditional plating in which a sample is spread on a nutrient plate and the number of yeast colonies that grow indicates the total number of live yeast cells present.

All these methods have shortcomings when applied to practical systems. Partly for this reason, and partly because it is a brutally complex and even chaotic system, the crucial connection between yeast growth and beer flavor is a somewhat confused field. Generally, however, flavor compounds appear in the medium roughly parallel with yeast growth, and perturbations that affect growth inevitably cause perturbations in the flavor profile. Also, it may be assumed that the entire metabolic composition of yeast, especially small molecules, is potentially available for leakage into the medium, and the yeast cell is hence able to contribute to beer an extraordinary spectrum of compounds that vary substantially in their flavor significance. Brewers therefore adhere rigidly to consistent brewing practices as a foundation for producing beers of satisfactory flavor quality. Chief among these practices are control of wort aeration, fermentation temperature and yeast quality and pitching rate, all of which impinge directly on yeast growth (rate and amount).

Brewing yeasts utilize fermentable sugars (glucose, fructose, maltose and maltotriose, sequentially in that order) as a source of energy (ATP) by the EMP pathway or glycolysis (Figure 11.2); glycolysis begins with formation of glucose-6-phosphate from glucose and ATP, and ends with pyruvate. This pathway is responsible for the three major products of fermentation: ethanol, carbon dioxide and glycerol. These products are not intermediates of the pathway itself but are produced when an essential metabolite, NADH, produced by oxidation (and phosphorylation) of glyceraldehyde-3-phosphate at step 6 in glycolysis, is reoxidized to NAD^+ (Figure 11.3). Without these reactions, which assure a continuing supply of NAD^+, glycolysis must cease. Glycerol has a sweet character and a soothing effect on harsh flavors; its consequence for good beer flavor perhaps deserves more study. Alcohol has a warming quality and helps signal richness and fullness. Carbon dioxide is crucial to the mouthfeel of beer being responsible for the "snap" or crispness of beers, especially light lagers served cold.

The amount of these compounds produced depends on the concentration of sugars present in wort; thus, using common rules of thumb, brewers can predict the yield of alcohol from the original gravity of wort or (conversely) can estimate the wort original gravity knowing the alcohol content. Therefore, the amount of raw materials (malt and adjunct) used to make the wort, malt diastatic power and the mash-temperature profile, each has a central role to play in beer flavor because these factors determine the total sugar in wort, the proportion of that sugar that is fermentable and the amount of other nutrients present, such as amino acids, vitamins and minerals. The primary flavor impact of yeast lies in the formation of a broad spectrum of flavor compounds present in small amounts that arise as products from metabolic pathways that lead to yeast growth (anabolism). Many of these materials are literally the end-products of metabolism that are of no further use to yeast; some simply leak from the metabolic pathway because they occur in excess, and some are products of metabolism that are degraded to make them less toxic to the cell and/or more easily excreted. In general, therefore, there is a positive correlation between yeast growth (requiring more metabolism) and formation of end-metabolites (flavor compounds). However, the correlation that is more to the point is between production of flavor compounds and metabolic flux (similar to, but not the same as, yeast growth). Thus, in a brewery fermentation, the rate of fermentation can be hugely affected by temperature, pitching rate and wort-dissolved oxygen without exactly parallel changes in yeast growth (rate or amount); there is therefore a disconnect between metabolic flux (catabolism, leading to flavor compounds) and synthesis of cell mass (anabolism). This accounts for the

Yeast 119

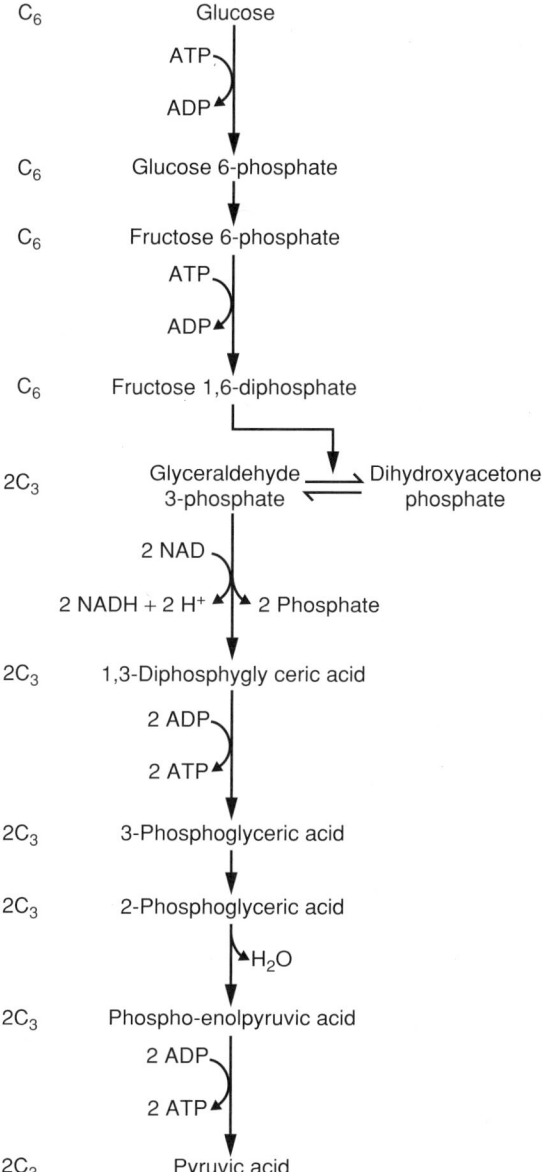

Figure 11.2. The Emden–Meyerhof–Parnas pathway of glycolysis

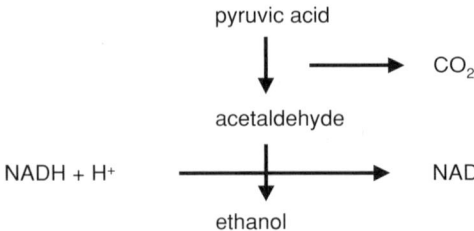

Figure 11.3. Replenishing reducing power by formation of ethanol

Assessing the Quality of Yeast

Yeast needs to be alive (high-viability) but also fit and healthy (high-vitality). The most common test for viability uses methylene blue: viable yeast decolorizes it, dead cells do not, and cells in each category are enumerated using a hemocytometer. Alternatively, a diluted suspension of yeast is applied to a microscope slide incorporating a layer of nutrient. After incubation cells can be distinguished as either giving microcolonies (viable) or not (dead).

There is no agreed technique for the assessment of vitality. Suggestions have been the measurement of glycogen, of sterols and of the rate at which yeast consumes oxygen. Probably, most favored is the assessment of acidification power: yeast is fed glucose and the extent to which the pH drops in response is an index of metabolic power.

fact that the same beer cannot be made at any fermentation temperature with any pitching rate or any wort oxygenation even with wort of identical composition. For any given product, brewers maintain a balance between yeast growth and metabolic flux that yields the fermentation rate/yeast-growth rate and, hence, product-flavor profile they require. Attempts to use high-gravity brewing (HGB) were at first frustrated by the inability to achieve the appropriate balance of these factors; though HGB will never make the same product brewed at ordinary gravity, the advantages of the HGB technology are such that brewer's are willing to accept a "good-enough match" for the standard product. This has been done by using very high pitching rates in HGB, greater wort aeration and sometimes temperature changes to maintain a satisfactory balance between yeast growth and metabolic flux. Two analogies are helpful. (1) When building a house, there must always be more building materials available than strictly necessary to assure there is no shortage at the job site; such excess is available for theft. Similarly, in growing yeast mass, catabolism must always produce more intermediates (building material) than is strictly necessary for yeast growth to assure there is no shortage for growth; this excess is available for leakage as flavor compounds. (2) A pipeline might be reasonably leak-proof when under low pressure, but may leak appreciable if the pressure is increased. The same is true for a metabolic pipeline or pathway: at low temperature and low speed it is relatively leak-proof, but leaks more at high temperature and speed. Thus, we might expect a high rate of fermentation and large yeast crop to release more metabolites (flavor compounds) than a low rate of fermentation even if the same amount of yeast is ultimately grown. Thus, lager yeast gives a more subtle or less bold flavor profile to beers, not so much because it is a lager yeast, but because yeast can be used at low fermentation temperature (low "pressure in the pipe" and so low release of flavor compounds). Lager yeast gives an ale-flavor profile when used at ale temperatures; if we could use ale yeast at lager temperature, doubtless it would produce a lager flavor profile.

The metabolic pathways by which individual components of beer flavor are produced by yeast are recorded in brewing textbooks. They comprise alcohols, acids, esters, aldehydes and ketones, etc. It is important to note that, in all cases, yeast *strain* is a central determinant of the end-products of metabolism that remain in the medium; therefore, appropriate choice of strain is always important and a first consideration in flavor control.

2,3-Butane-dione (diacetyl) and 2,3-pentane-dione (vicinal diketones or VDKs—Figure 6.1) cause an unwelcome buttery or butterscotch flavor in beers. They are a good example of metabolic intermediates that leak from pathways, particularly when fermentation is rapid. α-aceto-lactate and α-aceto-hydroxy-butyrate are common early intermediates in the synthesis of the amino acids valine/leucine and iso-leucine/threonine, respectively. In the medium, these precursors spontaneously (nonenzymically, copper and iron are involved) undergo oxidative decarboxylation to yield 2,3-butane-dione (diacetyl) and 2,3-pentane-dione. This is the rate-limiting reaction or slow step in diacetyl formation; thus, beer at one stage can be free of VDKs and later (even much later, e.g., after packaging) VDKs can arise. Brewers therefore measure VDKs *plus* VDK precursors to detect the full potential for this flavor defect. Yeast is able to remetabolize VDKs by taking them up and reducing them with NADH (so recovering NAD^+) to produce harmless products (diols). This is the basis of traditional secondary fermentation practices, e.g., lagering and krauesening. The "diacetyl rest" is a technique by which the fermentation temperature is maintained, or even increased and held, after primary fermentation is complete, until total VDK levels are satisfactorily low. The beer is then cooled. Production of VDKs is promoted by conditions that accelerate growth, especially in the early stages of fermentation and if certain amino acids (valine, leucine, isoleucine and threonine) are wanting. Diacetyl also arises from bacteria (see Chapter 6).

The most common ester of beer, by far, is ethyl acetate because the most common reactants available are ethanol and acetic acid. Acetic acid is readily available in active form as acetyl-coenzyme-A, which is a central starting point for anabolism (yeast growth); it derives from pyruvate by oxidative decarboxylation using NAD^+. Because other alcohols and other coenzyme-A's are also present, many other esters also occur; however, the esters of beer are, almost entirely either ethyl esters or acetate esters. They give fruity to solvent-like aromas. Esters arise as a result of the action of the enzyme alcohol acyl transferase on an alcohol *plus* an acyl-CoA. Ester control therefore is either through the availability of acyl-CoAs or the activity of the enzyme. In practice, this might be a distinction without a difference. Esters are probably a good example of an overflow mechanism for excess acyl-CoAs (e.g., as lipid synthesis slows) or a detoxification mechanism for toxic alcohol(s) (e.g., as they reach high levels). Therefore, esters tend to accumulate most if growth is restricted in some way and in high-gravity beers. Indeed, control of esters was a major accomplishment in making HGB practical. Oxygen in wort is a crucial factor in control of esters; less aeration than normal leads to higher ester levels because less than normal growth ensues.

Brewing yeasts have a well-developed mechanism for the decarboxylation of α-oxo-acids to aldehydes and then reduction of aldehydes to alcohols. These reactions are a rich source of many aldehydes and alcohols in beers. The reaction might be an example of detoxification mechanisms for α-oxo-acids enabling them to be more readily excreted into the medium. However, there is also a vital biochemical justification for the reactions: alcohol dehydrogenases reoxidize NADH to NAD^+ as they reduce aldehydes to alcohols. This gain in oxidizing power is crucial for the continuation of yeast's life under anaerobic conditions. The range of aldehydes and alcohols formed, then, depends of the spectrum of available α-oxo-acids. The primary α-oxo-acid is, of course, pyruvate, two molecules of which arise from each molecule of glucose fermented. Pyruvate yields ethan*al* (acetaldehyde) and ethan*ol* by this pathway. The most potent additional source of α-oxo-acids is amino acid metabolism, from either (1) the synthesis of amino acid precursors (i.e., α-oxo-acids) from carbohydrate metabolism or (2) the deamination of amino acids available in the wort to yield α-oxo-acids (Figure 11.4). Thus, the so-called α-oxo-acid pool within each yeast cell is charged from these two separate but related sources. Then, material can be withdrawn from the "pool" (1) to make required amino acids or (2) if in excess or if not required, to be excreted as higher (or "fusel") alcohols. Wort amino acids are not incorporated directly into yeast protein but are all deaminated. The valuable nitrogen (amino group) so freed is transferred by transamination to an α-oxo-acid to make an amino acid required for protein synthesis (this

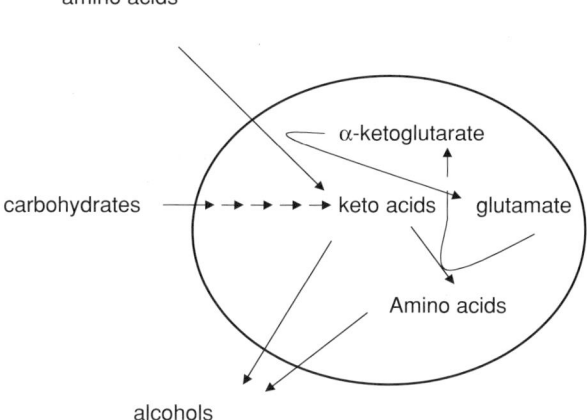

Figure 11.4. Fundamentals of nitrogen matabolism in yeast.

at least is the practical outcome). In this way the α-oxo-acid residue of wort amino acids become available for formation of higher alcohols; thus, the amino acid composition of wort (a more exact statement than wort FAN) that arises from malt can directly affect beer flavor. However, the yeast is also making α-oxo-acids that represent the protein content of the cell; this ameliorates the influence of the amino acid composition of wort on beer-flavor alcohols. Generally, we suppose that roughly equal amounts of higher alcohols arise from the two pathways identified, and that the catabolism (breakdown) of wort amino acids dominates in the early stages of fermentation and anabolic reaction (synthesis of α-oxo-acids) dominates in the later stages. Control of higher alcohol formation depends on control of yeast growth; factors that increase the amount of yeast crop (e.g., more oxygen in wort or higher wort nitrogen), or that accelerate the rate of yeast growth (or metabolic flux, e.g., fermentation temperature) or the amount of sugar to be metabolized (e.g., wort gravity especially in HGB), all tend to promote formation of higher alcohols. Ale yeasts make more higher alcohols than lager yeasts, but this is likely a function of fermentation temperature and possibly oxygen access (and so rate and amount of growth), rather than any fundamental biochemical difference between the two kinds of yeast.

Sulfur compounds, such as H_2S (hydrogen sulfide), SO_2 (sulfite) and DMS contribute significantly to beer flavor and, at adequately low concentrations, help define "beery" character. Because they have low flavor thresholds (can be detected by consumers in high dilution) their flavor contribution is out of proportion to their low concentration in beer, and an excess is unacceptable. Wort provides yeast with organic sulfur-containing compounds such as amino acids (cystine/cysteine and methionine), vitamins (biotin and thiamine) and metabolic intermediates (SMM) that are derived from malt and hence determined by malt modification, malt/adjunct ratio and wort original gravity. Pick-up of sulfur from malt and/or hops sulfured during kilning and the addition of KMS in the kettle (to control, e.g., color; see Chapter 3) can also influence the sulfur composition of wort and, especially if elemental sulfur enters wort, the flavor of beer. Inorganic sulfur, as sulfate ion in wort, is mostly derived from brewing water along with any additions of gypsum or Burton salts. These are the sources of beer sulfur compounds. In the absence of organic sulfur compounds, yeast can synthesize its requirements from sulfate; however, in fermentation of wort, yeast best utilizes methionine then cysteine to make such important metabolic cogs as Coenzyme A, TPP and S-adenosyl-methionine. Yeast directly incorporates sulfur-containing vitamins, if available, from the wort. These organic forms of sulfur spare the alternative energy-intensive mechanism, i.e., the reduction of sulfate to sulfite (source of sulfitic character in beer) and thence to H_2S (sulfidic character). Release of these compounds tends to happen in the later stages of fermentation after organic sources of sulfur have been exhausted.

DMS gives a cooked corn or veggie-fecal character to wort and beer, but in strictly controlled low amounts it is a positive character in some lager beers. DMS arises as a thermal breakdown product of SMM, which is a metabolic intermediate involved in many one-carbon transfers in living tissue. SMM arises as a result of malt modification and thus less-well modified malt yields less of this DMS precursor. SMM can break down at two points in brewing: malt kilning and wort boiling. If SMM is completely destroyed in kilning and the resulting DMS carried away in the airflow, none survives to break down in boiling; thus, highly kilned malts (e.g., traditional British ale malts and all specialty malts) are low in DMS production. Malts kilned at lower temperatures and low airflows are likely to allow SMM to survive into the wort where it breaks down during boiling. Intense boiling with generous evolution of steam drives off the DMS formed. Any SMM that survives boiling can break down in the whirlpool before cooling and yield a DMS-containing wort. Yeast also has a role to play in the DMS story. First, the vigorous evolution of CO_2 can help to carry off very volatile compounds, including H_2S and DMS. This happens particularly during the first 24 to 30 hours of a krauesen fermentation, for example, before the secondary fermenter is sealed to trap the CO_2 formed (for carbonation). Second, DMSO is formed alongside DMS in SMM breakdown; yeast readily reduces some this material, if present, to DMS.

Many brewers firmly hold that a length of copper pipe or other contact between wort and copper (e.g., the heat-exchange surface of a calandria) is hugely beneficial to minimize H_2S formation. It is assumed that the H_2S is rendered insoluble as copper sulfide. Generally, brewers believe that an H_2S problem, if not taken care of promptly by, e.g., gas flushing (a relatively simple procedure), will become a mercaptan problem that cannot be solved by any simple means.

Factors That Impact on Levels of Flavor Compounds Produced by Yeast

1. Higher alcohols
   Most significant = yeast strain, with ale strains producing more than lager yeasts
   Overoxygenation → increased production
   Increase in fermentation temperature → increased production
   Increase in fermenter pressure → decreased production
   (Factors causing increased yeast growth → increased production)
   Insufficient assimilable N (e.g., high sugar usage) → increased production (biosynthetic pathway)
   Excess assimilable N → increased production (Ehrlich pathway)
2. Esters
   Exponential increase in esters in relation to pitching gravity of wort (> 15°Plato)
   Strains differ in ester-producing capability
   Increase in fermenter pressure → decreased production
   (Factors causing increased yeast growth, e.g., high oxygen, lessen ester production—acetyl CoA diverted to cell biomass production rather than to esterification of higher alcohols)
   "Dirty worts" (high lipids) → decreased production
   Higher C:N ratio → increased production
   Increase in fermenter pressure → decreased production
3. Acetaldehyde
   Contaminator with *Zymomonas* can be a significant source
   Increase in fermenter pressure → increased production
   Factors inhibiting yeast activity → increased production
   Premature separation of yeast from beer → increased production
4. Vicinal diketones
   Contaminator with *Pediococcus* and *Lactobacillus* can be a significant source
   Factors inhibiting yeast activity → increased production
   Premature separation of yeast from beer → increased production
   Increase in fermentation temperature → decreased production
   Increase in pitching rate → decreased production
   Insufficient oxygen → increased production
   Insufficient free amino nitrogen → increased production
5. Hydrogen sulfide
   Yeast strains differ substantially in production

126 Chapter 11

Poor yeast vigor → increased production
Deficiency in pantothenate or vitamin B_6 → increased production
Shortage of zinc → increased production
Dirty worts → increased production
Copper → decreased production

6. Short- and medium-chain fatty acids
Lager strains produce more than ale strains
Increased oxygen → decreased production
Dirty worts → decreased production
Increased C:N ratio → decreased production
Increased pressure → increased production

Factors Impacting DMS Levels

Grist	1. Malted barley is only significant grist source of DMS precursors
	2. High N barleys give higher SMM—therefore, more SMM in six-row malts
	3. Increasing vigor of barley (during storage) gives increased SMM potential
	4. Increased embryo growth during modification gives increased SMM. Therefore, gibberellic acid promotes SMM and potassium bromate (and other rootlet inhibitors) suppresses
	5. Increased kilning temperature leads to reduced SMM. At higher temperatures, SMM is partially converted to DMSO. Also produced is methionine sulfoxide (MetSO; see later). Therefore, ale malts contain more DMSO and MetSO and less SMM than lager malt
Sweet wort production	1. Infusion mashing temperatures insufficiently high to degrade much SMM (see below). SMM and DMSO extracted. Decoction mashing will cause SMM degradation to DMS in the boiling segments
	2. Contaminating *Enterobacter* (e.g., *Obesumbacterium*) reduces DMSO to DMS
Boiling	1. SMM half-life at 100°C is 38 minutes. Every 6°C decrease in temperature leads to a doubling of half-life.
	2. Vigor of boil impacts volatilization of DMS released from SMM
	3. In insulated whirlpools, temperature is high enough to allow SMM to be degraded, but nonturbulent conditions means that most DMS released lingers

Hops/hop products	Small quantities of DMS in hop oil
Yeast and fermentation	1. Yeast reduces DMSO to DMS
	2. Yeast strains differ in this capability
	3. DMSO reduction inhibited by MetSO
	4. When yeast is N limited, it reduces more DMSO
	5. More DMSO to DMS conversion at lower fermentation temperatures
	6. Major volatilization of DMS with evolved CO_2. Depends on shape of vessel
	7. Disproportionately more DMSO reduction at higher gravities
	8. Higher pitching wort pH leads to more DMS production by yeast
Conditioning	Prolonged contact of yeast with beer will allow yeast to reduce DMSO to DMS
Filtration and stabilization	
Packaging	
Final product	Perception of DMS can be masked by phenylethanol

Flavors in Conflict

The perceived flavor of a beer is the net impact of a myriad of individual substances. Some have similar flavors and reinforce one another in impact. In other instances there seems to be a competition between flavors in ways that are not understood. A well-publicized example is how nitrogen gas suppresses hoppy aroma. Less well known is that of phenyl ethanol and phenyl ethyl acetate interfering with the perception of DMS. Thus in beers with low levels of phenyl ethanol or its acetate ester, DMS will be more readily perceptible, and vice versa.

Genetic Modification of Yeast

Should brewers become more tolerant of the genetic modification of yeast a number of opportunities will arise (read construct and rationale). Traditionalists will champion "alternative strategy."

Construct	Rationale	Alternative strategy
Flocculation factor (*FLO1* gene)	Promotion of flocculation to aid solid–liquid separation post fermentation	Centrifugation
Zymocin (killer toxin)	Killing off wild yeast and bacterial contaminants	Attention to plant hygiene
Glucoamylase (*DEX* or *STA2* genes)	Increased wort fermentability by eliminating dextrins	Addition of exogenous enzyme or extracts of very lightly kilned, extensively modified malt to fermenter
β–Glucanase	Continues the degradation of cereal glucans to avoid filtration problems	Selection of better and more homogeneously modified malts; low-temperature mash stands; addition of exogenous enzymes
Acetolactate decarboxylase (*ALDC* gene)	Converts precursor of diacetyl directly to less flavor-potent acetoin	Good fermentation practices: krausening (addition of a little freshly fermenting wort late in fermentation)
Elimination of sulphite reductase	Blocking enzyme that synthesizes hydrogen sulphide	Vigorous fermentation with healthy yeast
Elimination of sulphoxide reductase	Blocking enzyme that synthesizes DMS	DMS levels are determined by diverse factors, and not all of it is produced by yeast

Factors Impacting Extract Yield and Fermentability

Grist	1. Lower protein barley yields higher starch levels
	2. Two-row barleys have bigger corns and therefore more starch than two-row Six-row barleys have more nitrogen and therefore more enzyme potential for use in conjunction with starchy adjuncts devoid of enzymes
	3. Homogeneously modified malt, important to allow uniformity in milling
	4. Syrups and sugars added direct to kettle—"wort extenders"—therefore, allowing increased production in breweries lacking mashing capacity. Apart from barley syrups, these preparations do not supply assimilable nitrogen
Sweet wort production	1. Malt starch gelatinizes at 60–62°C, so 65°C is classic conversion temperature
	2. Rice, corn, sorghum starches must all be cooked
	3. α–Amylase seldom in limiting quantities (highly active, high thermotolerance). β-Amylase and limit dextrinase are less heat tolerant—but in 100% malt grist survive sufficiently. Will be increasingly likely to be limiting as malt is diluted with starchy adjuncts in mash tun
	4. Limit dextrinase is "limiting" because it is inhibited by endogenous materials from malt. Lowering pH to 5.1 releases more limit dextrinase, thus increasing fermentability
	5. At higher mashing temperatures (e.g., 72°C), β-amylase and limit dextrinase are rapidly destroyed; so high-dextrin (low-fermentable) worts produced
	6. FAN) produced in mashing should allow 140—150 mg FAN per liter for wort at 10° Plato
	7. Higher extractability of all malt components through hammer milling/mash filter. This includes starch-degrading enzymes and therefore fermentability as well as overall Extract.
	8. Parti-gyling: separate collection of strong and weak worts. One of the techniques for achieving HGB
	9. The extract in last runnings has a different composition to that earlier in lautering—less-assimilable sugars and FAN, more phenolics and ash materials
Boiling	1. Addition of wort extenders (see above). Brewer can "dial" for fermentability of these worts
	2. Concentration of wort—cf. HGB
	3. Enzyme inactivation
Hops/hop products	
Yeast and fermentation	1. Yeast strains differ in their ability to deal with carbohydrates
	2. Yeast must receive sufficient oxygen and zinc

	3. Pitching of 1 million cells/ml/degree Plato of viable yeast
	4. Cloudier worts give more vigorous fermentation
	5. Addition of exogenous glucoamylase allows complete conversion of dextrins to fermentable glucose
Conditioning	Addition of priming sugar either for sweetness or for "natural" conditioning
Filtration and stabilization	Addition of deaerated water of identical salt content to beer for diluting to desired strength
Packaging	
Final product	

|12|

Oxygen

The atmosphere of this planet is 20% oxygen; unfortunately, too much oxygen in the wrong place and at the wrong time can be devastating to beer quality. While anaerobiosis gives brewers an important leg-up in microbiological terms, historically the most important problem of beer stability (see Chapter 6), that is about the only advantage of it. These days, when fresh-flavor shelf-life is the main focus of beer stability, brewers must spend an inordinate amount of time and money fighting oxygen.

Oxygen becomes an increasing problem for brewers as the product advances through the brewery and reaches a peak after the yeast is removed. Bright beer in the cellars must be assiduously protected from oxygen pick-up, e.g., during transfer to the packaging hall; this is because oxygen dissolves readily in cold aqueous systems in which the level of the gas is already low and if turbulence happens (below). This describes cellar beer as it is moved in pipes, pumps and tanks about the brewery. In bulk cellar beer, the lower limit of present technology is about 0.2 parts per billion of dissolved oxygen.

Oxygen can first enter the brewing process at mash-in from air entrained in the grist, dissolved in mashing water and picked up during the agitation and transfers of the brewhouse. It is probably at this stage that oxygen is most reactive in brewing processes because the variety and concentration of potential reactants, the catalysts that might promote reactions (enzymes and

Measurement of Oxygen

The operation of the oxygen electrode is shown in Figure 12.1. At the cathode, oxygen is reduced by electrons to produce water. Those electrons are supplied by the reaction of chloride from the electrolyte (potassium chloride) with the silver anode. The more oxygen present, the more current flows.

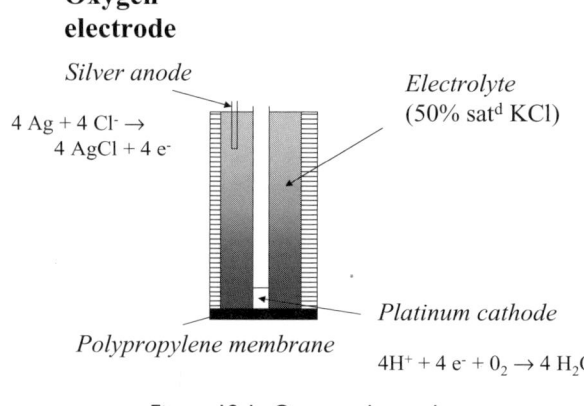

Figure 12.1. Oxygen electrode.

metal ions) and the driving force of most chemical reactions (heat) are all at their highest level. While there is no detectable dissolved oxygen in the mash, the turnover or pass-through of oxygen could be substantial because of the reactivity of oxygen. It is not surprising therefore that in recent times some brewers have turned their attention to limiting oxygen access at this stage in so-called "anaerobic" brewhouses. Nevertheless, this is perhaps a marginal choice and will be a productive approach to limiting the damaging effects of oxygen only if oxygen is first thoroughly controlled downstream (especially in the cellars).

The importance of oxygen to yeast growth is mentioned elsewhere (see Chapter 11). Wort aeration or, for higher level of dissolved O_2, oxygenation, is the only example of deliberate addition of oxygen in brewing processes and is strictly controlled to the required level. Oxygen aids the yeast to form unsaturated fatty acids that are necessary for construction of cell membranes;

Elimination of Oxygen in the Brewhouse

Although the evidence is not entirely convincing that limiting oxygen uptake in the brewhouse manifestly benefits flavor stability, there are many brewers who strive to minimize air ingress at all stages in the brewery. It certainly makes sense to take sensible precautions (of the type listed in the box "Factor impacting on flavor stability"); however, to go to extremes such as operating the entire brewhouse under an inert atmosphere is surely overkill. It may even be detrimental—e.g., to haze stability. Intermediate between the two extremes (doing nothing to prevent air ingress or the oxygen-free brewhouse) are precautions such as mashing in with deaerated water or purging the milled grist with nitrogen. To get an idea of the relative worth of each of these, the water may contribute 10 g of oxygen per ton of malt, whereas the grist may have trapped within it some 600 g of oxygen per ton.

this is assisted by unsaturated fatty acids and sterols derived in wort from malt and utilized by yeast, with the expenditure of energy, to form acyl-CoAs. This added oxygen is exhausted early in fermentation and the process is thereafter anaerobic. Yeast is an efficient scavenger for entrained air and so oxygen pick-up in beer that contains yeast is much less damaging than in bright beer, though both are bad practices.

Oxygen to Yeast or to Wort?

Oxygen is added to the wort as a nutrient for the yeast. There is some reaction of oxygen with wort constituents and some believe that this is not only wasteful but also detrimental to the flavor stability of beer. Accordingly, it has been argued that it makes more sense to oxygenate the yeast immediately before pitching *per se*, rather than the wort. Provided the physical challenges inherent in trying to get oxygen to all of the cells in a thick slurry are overcome, it is certainly the case that improved fermentation control can be achieved by pitching defined quantities of oxygenated yeast.

Must We Give the Yeast Oxygen?

Oxygen is required by the yeast to produce the lipid molecules (sterols and unsaturated fatty acids) that are significant components of its membranes. If sufficient quantities of these materials are directly provided to the yeast then the addition of oxygen is superfluous. No wort contains enough lipid; however, it is possible to supplement with lipid, and spent grain pressings have been suggested as one potential source. Such an addition will also act as an antifoam. As yet, no brewer adopts such a practice.

Lipids and polyphenols are extracted from malt in mashing and are the main substrates available for oxidation.

The lipids of malt are of many different kinds (comprising some 3%–4% of malt dry weight) and arise mainly from the living tissue of the barley, the embryo and aleurone layer, with triglycerides dominating. Most (90% or more) of this complex lipid mixture remains in the spent grain and only the more soluble free fatty acids, some likely formed from triglycerides by the action of malt lipases during mashing, plus some triglycerides and phospho-/glyco-lipids are extracted into wort. The amount present is greater in all-malt worts compared to those made with low-lipid adjuncts. Extraction also increases with malt modification (e.g., because of growth of the plumule) and any factors that make lipids more likely to dissolve, e.g., fine milling, high mash or sparge temperature, vigorous agitation or raking. Any other

Enzymes from Barley That React with Polyphenols

Enzyme	Mode of action	Products
Polyphenol oxidase	Reacts polyphenol with oxygen	Quinones
Peroxidase	Reacts polyphenol with hydrogen peroxide	Quinones

factors that tend to maximize extract yield and/or lead to hazy worts add to the lipid content of wort. Wort lipids, especially fatty acids, might be taken up by yeast during fermentation and so indirectly influence beer flavor by affecting yeast growth, as noted above. Also, fatty acids can be esterified to the ethyl-fatty acid by yeast action and any aldehydes that arise from lipids in mashing can be reduced during fermentation to the corresponding (harmless) alcohol. Some products might be volatile in kettle boiling and others separate with the trub. As a result, although a wide array of long-chain fatty acids and related esters and alcohols can be found in beer, the levels are very low (below flavor threshold) and their direct effect on beer flavor is minimal.

The most consequential reaction of lipids, especially unsaturated fatty acids, is with oxygen. This is a well-known reaction of fats, generally referred to as rancidity, and results in the formation of aldehydes. Brewers generally acknowledge that long-chain unsaturated aldehydes are the origin of stale flavor in beer. These can arise from long-chain unsaturated fatty acids such as the C-18 fatty acids: oleic acid (1=), linoleic acid (2=, which is about 80% of such acids in wort) and linolenic acid (3=) by reaction with oxygen. The most famous of such aldehydes is 2-trans-non-ene-al, which was once thought to be the silver bullet that defined stale flavor. The more highly unsaturated fatty acids are more reactive with oxygen, and linolenic acid (a di-ene and in highest concentration in wort) in particular, is likely to be effective. The enzyme LOX or lipoxygenase has been implicated in oxidation of malt lipids in mashing; however, the enzyme is substantially

Lipid-Degrading Enzymes from Malted Barley

Enzyme	Mode of action	Products
Lipase	Hydrolyzes ester linkages between fatty acids and glycerol	Fatty acids, monoglycerides, diglycerides, glycerol
Lipoxygenase	Catalyzes the reaction of polyunsaturated fatty acids with oxygen	Hydroperoxides
Hydroperoxide lyase	Splits hydroperoxides	Aldehydes
Hydroperoxide isomerase	Rearranges hydroperoxides	Epoxyhydroxy acids
Hydrase	Adds water to epoxyhydroxy acids	Trihydroxy unsaturated fatty acids

inactivated by kilning, is at an unsuitable pH and temperature in the mash and must scramble to find an essential substrate for action—like molecular oxygen.

As noted, the reaction of fatty acids with oxygen during mashing is likely to be much affected and ameliorated by downstream processes such as the kettle boil (e.g., lipids separate with trub) and during fermentation (e.g., the utilization and modification of fatty acids). It is the long-term (measured in months) reaction of lipids with oxygen in beer that affects beer flavor, causes loss of fresh beer flavor or is the genesis of stale flavor. This is often described as papery or cardboard, but many of the terms used to describe it are less pejorative such as malty, caramel, toasty; these are potentially positive characters that in the wrong place are, nevertheless, undesirable. As noted, it is unlikely that any of the staling aldehydes, formed in mashing, themselves form stale *beer* flavor. More likely, some intermediate of those reactions, likely bound to polypeptides, hop resins or as salts with divalent ions, are stable enough to survive boiling and soluble enough to survive into beer; there, their continued slow release and breakdown causes the flavor change associate with loss of fresh flavor. Tri-hydroxy-fatty acids, related to oleic, linoleic and linolenic acids, have been suggested as these precursors. Alternatively, auto-oxidation can occur over time; this requires oxygen radicals to be present, and metal ions especially copper and iron probably also participate. Oxygen itself is not particularly a reactive molecule, but by reduction (e.g., in beer) a superoxide radical (O_2^{-*}) can form; and from it the very reactive, though short lived, hydro-peroxide (HOO*) and hydroxyl (HO*) radicals can arise. Unsaturated fatty acids or their derivatives

Enzymes from Barley That Scavenge Active Forms of Oxygen

Enzyme	Mode of action	Products
Superoxide dismutase	Eliminates superoxide radicals two at a time	Equal quantities of peroxide and ground-state oxygen
Catalase	Eliminates peroxide two molecules at a time	One molecule of ground-state oxygen and two molecules of water
Peroxidase	Reacts polyphenol with hydrogen peroxide	Quinones

that survive into beer are then prone to a self-perpetuating oxidation with the formation of long-chain aldehydes with low-flavor thresholds and, so, stale flavor.

PHENOLIC ACIDS

The benzoic acid series and cinnamic acid series of phenolic acids, adding up to a few parts per million in beers, form part of the acidic background of beer; they are not otherwise characteristically flavorful. Much more consequential are the polyphenols because they participate in precipitation of proteins at all stages of the brewing process, including importantly, haze-forming reactions in beers (see Chapter 5), and they are a cause of astringent mouthfeel characteristics of some beers. They also react with oxygen forming phlobaphenes that add color to beer and account for the slight darkening of pale beers with age. This reaction however also has antioxidant potential that could protect beer from oxidation by using up oxygen and quenching radical-driven reactions. However, the oxidized polyphenol might itself act as a donor, or oxidant molecule, under some circumstances, especially the presence of metal ions (again copper and iron), because its oxidized form contains a structure not unlike that of reductones, i.e., an unsaturated α-dicarbonyl. Reductones contain the general structure $HOCH_2COCHO$ (reductone itself); they react with oxygen to give dehydro forms that are (again) α-di-carbonyls. Ascorbic acid (vitamin C) is a classic example of a reductone and is sometimes added to beer to conserve fresh flavor. Once again, however, the oxidized molecule could participate in oxidizing reactions and so act as an oxygen carrier. There is a long-standing understanding by brewers that there is a general correlation between oxidation of polyphenols, change of beer color, lower reducing power (redox potential) of beer, formation of haze and loss of fresh flavor; brewers assume that these reactions are interconnected. Indeed, the Strecker degradation (see Chapter 3), between α-dicarbonyls and amino compounds, provides an opportunity to form aldehydes that might influence flavor. Reactions such as this might also explain the suggestion that melanoidins (products of the Maillard reaction that also can involve the Strecker degradation) are involved in formation of aldehydes, though brewers observe that dark beers are intrinsically more stable to flavor change by oxidation than pale beers.

Sulfur dioxide (SO_2) is a powerful antioxidant, though the levels in beer permitted by regulation (10 mg/l in the USA) are generally insufficient to confer flavor stability; it forms addition compounds with aldehydes and so in beer is mostly bound SO_2. SO_2 arises in beer by yeast action and is a variable depending on yeast strain, or arises by addition of KMS in the kettle or post-fermentation. It is an effective flavor preservative at levels somewhat below its flavor threshold of about 25 mg/l.

Pathways to Staling Substances

The most frequently cited pathway by which staling aldehydes are produced is through the oxidation of unsaturated fatty acid to form hydroperoxides, which are subsequently transformed enzymatically and then nonenzymatically to the lower molecular weight unsaturated aldehyde (E)-2-nonenal that has pronounced cardboard character.

However, it is naive to believe that nonenal is the only contributor to stale character—diverse unpleasant compounds containing the carbonyl (C=O) group are found in beer and they are produced via other pathways.

Alcohols in beer can be converted to their equivalent aldehydes. This is a reaction catalyzed by melanoidins, substances often referred to as antioxidants but which can have undesirable impacts also.

Iso-α-acids are oxidized with the formation of carbonyl substances from their side chains. The reduced derivatives do not do this.

The Strecker degradation comprises a reaction between an amino acid and an α-dicarbonyl compound, such as the intermediates in browning reactions. The amino acid is converted into an aldehyde with one fewer carbon atom. The reaction is believed to be catalyzed by certain polyphenols, showing again that some classes of substances may have both beneficial and adverse roles to play.

Different carbonyl-containing substances produced in the types of reaction listed above can react together in the "aldol condensation" to form larger, different carbonyl compounds; for example, (E)-2-nonenal can come from the reaction between acetaldehyde and heptanal. Proline, which is abundant in beer, can catalyze such reactions.

Polyphenols as Pro- and Antioxidants

Polyphenols with hydroxyl groups at the 3′ and 4′ positions on the flavan ring (e.g., catechin) are antioxidants because they scavenge oxygen radicals. Those with an additional 5′ hydroxyl group (e.g., delphinidin) promote staling because they can reduce transition metal ions to their more potent lower valence forms, e.g.,

$$2Cu^{2+} + RH_2 \rightarrow 2Cu^+ + R$$

$$Cu^+ + O_2 \rightarrow Cu^{2+} + O_2^-$$

$$Cu^+ + O_2^- + 2H^+ \rightarrow Cu^{2+} + H_2O_2$$

$$Cu^+ + H_2O_2 \rightarrow Cu^{2+} + OH^- + OH^\bullet$$

O_2^- = superoxide, H_2O_2 = peroxide, $OH^\bullet$ = hydroxyl

Prediction of Flavor Stability of Beer

As per normal in brewing, there is an obsession with having tests, which can be applied to beer to predict how long its flavor life will hold up. In fact, the only merit in having tests for shelf-life is if they enable the brewer to respond in some way such that stability is enhanced, there being absolutely no benefit from knowing that a product in package is going to have a low flavor life (unless the brewer is prepared to withdraw such beer from trade or even prevent release to trade, which would be a prohibitively expensive option).

Worthwhile procedures might be applied in two ways:

(1) Those that can be applied in a process research context in order to establish raw materials, process conditions, etc. that will lead to enhanced product stability.
(2) Those that could be applied in a QA/QC set-up to indicate raw material, process and product status to enable the brewer to make adjustments in order to favor shelf-life.

To expedite the study of what is inherently (and preferably!) a long-term phenomenon, several researchers have developed forced ageing procedures. Typical regimes are, holding beer at either 60°C for 24 hours, or 37°C for 3 weeks, or 30°C for 4 weeks prior to evaluation by tasting. These are said to be good mimics for 6 months at 18°C. Another procedure involves shaking beer (to simulate transport—a factor insufficiently considered in the context of flavor life) followed by 4 days of storage at

40°C. It is claimed that this equates to 3 to 4 months at 20°C. Critics of these accelerated regimes say that the flavors obtained at the higher temperatures are different to those developing during natural storage.

As an alternative to tasting, some have advocated the chemical monitoring of species produced in forced ageing techniques. Thiobarbituric acid (TBA) has been used to measure carbonyl species produced on forced ageing, but TBA is not a very specific agent and preferentially reacts with malondialdehyde, which is but one of the breakdown products from unsaturated fatty acids. Another breakdown product is ethylene, which has also been cited as an indicator of staling potential.

Some promote the concept of "nonenal potential" for the assessment of oxidation in the brewhouse. Worts are heated to release nonenal, which is measured; higher levels of nonenal are said to relate to more extensive oxidation during wort production and worts with high nonenal potential are believed to proceed to beers with a greater propensity to staling.

Other species are sometimes measured directly as indices of staling. Notable amongst these is furfural, which is not believed to directly contribute to staling per se, but it is felt that it is a good yardstick for oxidation. Another way to assess overall oxidation in the process stream is the indicator time test, in which samples are incubated with 2,6-dichlorophenolindophenol (DCPIP). When DCPIP is in the reduced form it is colorless, but when it is oxidized it is blue. The extent of blue coloration (as assessed spectrophotometrically) is therefore an indication of the overall extent of oxidation that has occurred in the brewhouse.

Particular interest has been paid in recent years to the direct measurement of radicals in beer and its process stream. This is rationalized on the basis that it is the radical forms of oxygen that are key to flavor instability; therefore, tools to measure the level of radicals in beer should give the best possible indication of staling potential.

One approach is to measure chemiluminescence, both directly and after reaction of beer with the radical scavenger 2-methyl-6-phenyl-3,7-dihydroimidazo (1,2-a) pyrazin-3-one (CLA). This is allied to the use of 1,1-diphenyl-2-picryl-hydrazyl (DPPH) as a measure of reducing power in beer (DPPH being claimed to be a preferred alternative to DCPIP). It seems that DPPH correlates with polyphenol species, though not with SO_2, and that the value increases through brewhouse operations, but declines during fermentation.

Others advocate the measurement of radicals by the application of electron spin resonance technology (ESR). The endogenous antioxidant (EA) value is the time taken before an ESR signal is developed in an ageing test; the longer the lag, the greater the antioxidant potential of the sample. Among the interesting observations made using this approach, it was shown that the EA value is especially developed during fermentation (due to SO_2

production?) and that temperature of storage seemed to have a much bigger effect than level of oxygen on the development of signals due to radical species. This finding should be compared with those of others that raising temperature from 0°C, though 25°C, to 40°C has a disproportionate effect *inter alia* on the loss of iso-α-acids from beer and on the levels of furfural developed. It certainly does seem that the most important factor determining the rate of stale flavor development is temperature, albeit not all compounds increase in level with increased temperature.

A somewhat less-expensive technique, albeit one which is not yet satisfied by having robust instrumentation commercially available, is the measurement of redox potential. Such values give an overall indication of the oxidation–reduction status of a sample and would be expected, for instance, to give a more meaningful indication of oxidative damage when applied to beer than would the measurement of oxygen, because oxygen will be rapidly consumed in package, particularly during pasteurization.

Simpler colorimetric techniques for assessing oxidative damage include the use of iodine staining or the measurement of free thiol groups using 5,5′-dithiobis (2-nitrobenzoic acid) (DTNB).

Scavenging Corks

Oxygen can creep between the crown cork and lip of a bottle, leading to a substantial gas pick-up over time. One approach to prevent this is to use oxygen-scavenging crown corks. These employ metal-catalyzed oxidation of a polymer sandwiched between layers of a polymer such as PET.

Factors Impacting on Flavor Stability

Grist	1. More highly kilned pale malts contain less LOX
	2. Darker malts contain more antioxidant Maillard reaction products
	3. Nonbarley adjuncts (degermed rice and corn grits; sugars and syrups) lack staling precursors

Sweet wort production	1. Milling that preserves embryo tissue undamaged will not release LOX or lipids
	2. Milling under inert gas to avoid air ingress
	3. Purging of air from milled grist with nitrogen or carbon dioxide
	4. Mashing with deaerated water
	5. Use of a premasher
	6. Mashing-in at highest practical temperature to obviate LOX action
	7. Mashing at lower pH (<5.2) to prevent LOX
	8. Fewest number of transfers and pumping to minimize opportunity for air ingress*
	9. Turning off pumps when transfer complete
	10. Filling vessels from bottom*
	11. Avoid use of rousers until covered*
	12. Copper from copper vessels will promote production of damaging radicals*
	13. Inert gas as motor gas
	14. Good plant maintenance (e.g., correct leaking pumps)
Boiling	1. Vigorous boiling to purge volatiles
	2. Excessive boiling leads to production of "cooked" flavors
Hops/hop products	1. Reduced iso-α-acids more resistant to degradation to carbonyl compounds
	2. *Trans* isomers of iso-α-acids more prone to degradation to stale compounds: ratio of *cis:trans* is 2:1 for conventional boiling with hops or pellets, but 5:1 for isomerized extracts; so latter may offer more stability
Yeast and fermentation	1. Good yeast husbandry (pitching rates, viability, vitality) to promote scavenging of carbonyls
	2. Promotion of SO_2 production to bind staling carbonyls—to increase SO_2, increase sulfate supply to the yeast, increase wort clarity, increase oxygenation of wort, reduce pitching rate, reduce fermentation temperature
	3. Higher out of fermenter pHs preferable (see below)
Conditioning	
Filtration and stabilization	1. Use of low-iron filter aids (and other additions and process aids)
	2. Divert water used to precoat filters to drain
Packaging	1. Double evacuation, inert gases, tappers and jetters, undercover gassing and other low-air filling protocols
	2. Scavenger crown corks
Final product	1. Beer progressively more susceptible to staling as pH is lowered from 4.5
	2. Store and transport beer as cold as possible, but short of freezing

*Applies at other process stages also.

|13|

Water and Energy

Water and energy are connected in the brewery: energy is almost entirely used to heat, cool or evaporate water of aqueous systems, and the high specific heat of water assures that much energy is required for this. Both commodities have become increasingly expensive over the years, and in most recent times energy especially so. Conservation of energy and water and control of pollution have become watchwords of modern brewing. Most breweries use 4 to 6 hectoliters of water per hectoliter of beer produced, but some use much more, especially small breweries.

Water first affects brewers as soil moisture in the barley-growing districts. It arises as snowmelt water or as winter rains. Sufficient, but not excessive, soil moisture permits the soil to warm up quickly in Spring and be ready to be planted with seed. Spring rain, or in some areas irrigation, assures the barley plant grows to maturity and forms an adequate head in which the seeds fill properly; absence of appropriate rain during ripening and harvest contribute to a satisfactory barley crop. Barley for brewing use, among other qualities (see Chapter 8), must be of low moisture content suitable for prolonged (up to 15 months) storage. Moisture content and temperature are the variables that determine how long a lot of barley can be stored. Barley harvested at high moisture must be dried for storage.

After preparation for malting barley enters the steep, in which the grain is buried in water. Water for steeping must be cool (about 12°C), potable and free of excessive iron, sulfur or other noxious elements. A cool steep assures that water uptake is properly paced so that each kernel is evenly wetted; the process takes about 2 days. Warmer water accelerates the process but also promotes unevenness of water uptake. The husk rapidly takes up water and is regarded as a reservoir from which the grain can absorb water during, e.g., steeping air rests or in germination, and the water is siphoned to the micropyle through the ventral crease. Water enters the grain itself through the micropyle region close to the embryo where the testa is thin or absent. The embryo therefore also hydrates quite quickly. However, water penetration into the endosperm takes much more time, and it is here that "mealy" (rather than "steely") kernels are preferred. In a mealy endosperm water penetrates more rapidly and evenly from the embryo (proximal) end toward the distal end of the kernel, and more rapidly on the dorsal side than the ventral. Sufficient and even moisture uptake by the endosperm is required if sufficient and even modification of the endosperm is to follow during the germination phase of malting.

Maltsters wish to limit water use because water is expensive to acquire and to dispose of, especially in the heavily polluted form (steep water) in which it leaves the malt house. Steep water performs two main functions: (1) it washes the grain including removal of microbes and (2) provides the water for grain hydration. It is now quite common to separate these two functions and first put the grain through a barley washer in which it is aggressively and efficiently cleaned before it enters the steep proper. There is no reason why different conditions cannot be used in barley washing and barley steeping, e.g., agitation, aeration, lime cleaner and temperature. This saves water. In steeping, barley takes up water and begins to respire and the microbes associated with the grain also take up oxygen and produce CO_2. It is therefore necessary to assure a sufficient supply of air (oxygen) to prevent the grain from suffocating. The barley swells to nearly 1.5× original volume as it takes up water.

The objective of steeping is to assure that grain reaches the overall moisture content required for the malt being made; this is almost always in the range 42% (e.g., for regular pale malt of average modification made from a vigorous variety of barley) to say 48% (for dark malt or well-modified malt, especially one made from a less vigorous variety). The "steep-out" moisture of barley therefore is a major determinant of future malt quality and an harbinger of malting losses that might accrue; generally higher malting losses result from higher steep-out moistures. The end of steeping is signaled by the appearance of the "chit" or coleorhiza. Further exposure to water after chitting would drown the grain.

Water is occasionally added to grain during germination if the grain is falling behind in its development. This water almost certainly affects only the embryo. Water addition also counteracts the tendency of the grain bed to dry out under the influence of airflow in the germination vessel. The airflow is thoroughly humidified by drawing the air through sheets of water to clean, humidify and cool it (by latent heat of evaporation). But, as the grain bed tends to warm up during germination, the extra heat increases the water-carrying capacity of the air; the grain bed therefore tends to dry out. The husk, but more importantly the embryo, dries first, which tends to impede modification especially if it happens in the early stages of germination. Some drying (called "withering") might be desirable toward the end of germination when drying/inhibition of the respiring embryo might help to control malting loss and prepare the grain for kilning.

Kilning is the most energy-intensive operation in malting in which large volumes of air are moved and heated and a great weight of water evaporated. Maltsters cannot avoid the present high (and rising) cost of fuel and unfortunately there are few strategies available for greater fuel efficiency. Reusing the air off the lower kiln again on the upper kiln (a long-standing practice), and preheating incoming air with exhaust air are useful practices. Cogeneration burns fuel in a turbine to generate electricity that can be sold; the "waste" heat can then be used for kilning. Geothermal and solar sources are not used at the moment though these could be possible resources in some locations. Malts with 6% moisture are available and should be cheaper because removing the last few pounds of moisture is relatively expensive and perhaps unnecessary. Taking this idea to the extreme, making beer with green (unkilned) malt turns out to be inconvenient and produces poor beer. However, acceptable beer can be made from barley plus enzymes; this well known technology might have a future in the present energy climate, especially for new products (i.e., those that do not have to match existing flavor profiles).

The purpose of kilning is to evaporate the moisture and other unwanted volatiles present in the green malt and then toast the malt lightly to imbue, through the Maillard reaction primarily (see Chapter 3), the characteristic color and malty/biscuity/toasty flavors of malt. The vast bulk of the energy is required to lower the moisture content of malt 10-fold (say) from 45% to 4.5%. Kilns therefore are operated as evaporators and toasters by manipulating air flow volumes and temperatures. A two-floor kiln allows the air-off the lower floor, e.g., where the grain is being toasted, to be diluted with cool air and then be reused to evaporate water from malt on the upper kiln. The exit air, carrying a full load of moisture, still contains useful heat and is often used to preheat the fresh incoming air for energy conservation. In all cases of course dry air intake is preferred to humid air.

The evaporation of water at the early stages of kilning is a key event in the preservation of the enzymes of malt. The protective factor is the latent heat of water evaporation: as the water evaporates it cools the malt. Of course as the malt dries this effect is progressively less relevant. The practical manifestation of the latent heat of water evaporation is the difference between the temperature of the air entering a grain bed (the "air-on") and the temperature of the air leaving it (the "air-off"). This difference might be some

Water Regulations

In the United States water must satisfy the National Primary Drinking Water Regulations established by the Environmental Protection Agency. These are summarized in the first table. Additionally, there are National Secondary Drinking Water Regulations—see second table. The latter are merely guidelines, not enforceable by law.

Extract from the National Primary Drinking Water Regulations

Component	Maximum contaminant level goal	Maximum contaminant level (mg/l unless stated)	Potential health effects	Sources of contaminant
Cryptosporidium or *Giardia*	Zero	99–99.9% removal/ inactivation	Diarrhea, vomiting, cramps	Fecal waste
Legionella	Zero	Deemed to be controlled if Giardia is defeated	Legionnaire's disease	Multiplies in water heating systems
Coliforms (including *Escherichia coli*)	Zero	No more than 5% samples positive within a month	Indicator of presence of other potentially harmful bacteria	Coliforms naturally present in the environment; *E. coli* comes from fecal waste
Turbidity	n/a	<1 nephelometric turbidity unit.	General indicator of contamination, including by microbes	Soil runoff
Bromate	Zero	0.01	Risk of cancer	Byproduct of disinfection
Chlorine	4	4	Eye/nose irritation; stomach discomfort	Additive to control microbes
Chlorine dioxide	0.8	0.8	Anemia; nervous system effects	Additive to control microbes
Haloacetic acids (e.g., trichloracetic)		0.06	Risk of cancer	Byproduct of disinfection
Trihalomethanes		0.08	Liver, kidney or central nervous system ills, risk of cancer	Byproduct of disinfection

Component	Maximum contaminant level goal	Maximum contaminant level (mg/l unless stated)	Potential health effects	Sources of contaminant
Arsenic		0.05	Skin damage, circulation problems, risk of cancer	Erosion of natural deposits; runoff from glass and electronics production wastes
Asbestos	7 million fibers per liter	7 million fibers per liter	Benign intestinal polyps	Decay of asbestos cement in water mains; erosion of natural deposits
Copper	1.3	1.3	Gastrointestinal distress, liver or kidney damage	Corrosion of household plumbing systems; erosion of natural deposits
Fluoride	4	4	Bone disease	Additive to promote strong teeth; erosion of natural deposits
Lead	zero	0.015	Kidney problems; high blood pressure	Corrosion of household plumbing systems; erosion of natural deposits
Nitrate	10	10	Blue Baby syndrome	Runoff from fertilizer use, leaching from septic tanks, sewage, erosion of natural deposits
Nitrite	1	1	Blue Baby syndrome	Runoff from fertilizer use, leaching from septic tanks, sewage, erosion of natural deposits
Selenium	0.05	0.05	Hair or fingernail loss, circulatory problems, numbness in fingers and toes	Discharge from petroleum refineries, erosion of natural deposits, discharge from mines
Benzene	zero	0.005	Anemia; decrease in blood platelets; risk of cancer	Discharge from factories; leaching from gas storage tanks and landfills

Component	Maximum contaminant level goal	Maximum contaminant level (mg/l unless stated)	Potential health effects	Sources of contaminant
Carbon tetrachloride	zero	0.005	Liver problems; risk of cancer	Discharge from chemical plants and other industrial activities
Dinoseb	0.007	0.007	Reproductive difficulties	Runoff from herbicide use
Dioxin	zero	0.00000003	Reproductive difficulties, risk of cancer	Emissions from waste incineration and other combustion; discharge from chemical factories
Alpha particles	Zero	15 picoCuries per liter	Risk of cancer	Erosion of natural deposits
Beta particles and photon emitters	Zero	4 millirems per year	Risk of cancer	Decay of natural and man-made deposits

The full table can be found at http://www.epa.gov/safewater/mcl.html

National Secondary Drinking Water Regulations

Contaminant	Secondary standard
Aluminum	0.05–0.2 mg/l
Chloride	250 mg/l
Color	15 color units
Copper	1 mg/l
Corrosivity	Noncorrosive
Fluoride	2 mg/l
Foaming agents	0.5 mg/l
Iron	0.3 mg/l
Manganese	0.05 mg/l
Odor	3 threshold odor number
PH	6.5 – 8.5
Silver	0.1 mg/l
Sulfate	250 mg/l
Total dissolved solids	500 mg/l
Zinc	5 mg/l

These are nonenforceable guidelines regulating contaminants that may cause cosmetic effects (e.g., skin or tooth discoloration) or aesthetic effects (taste, odor, color). States may choose to adopt them as enforceable standards.

30°C with wet grain and shrink to nothing as the grain dries; that signals the end of drying.

Malt is hygroscopic and will take up moisture resulting in "slack" and inferior malt. This is especially true for milled malt. All malt products therefore should be stored in dry, cool, sanitary and dust-free conditions.

In the brewery water performs numerous functions e.g. in sanitation & steam raising. In this chapter we look only at water that enters the product.

MASHING

If all the water required to make wort were added to the grist in one batch, the size of brewhouse vessels would have to be much greater than they are in modern practice. It requires about 15 kg of milled malt to make one hectoliter of wort (or about 38 lb to make one barrel US) of ordinary specific gravity. However, a good deal more water must be put into the process to yield one hectoliter of wort because much water is lost in the spent grain and some water is evaporated during boiling.

Water is added to grist in two main batches: as mash water (usually one-third to one-half of the total water) and as sparge water (the remaining volume). The volume added as mash water determines mash thickness; a brewers' rule of thumb is that the water:grist ratio should not be less than 2.5:1 (2.5 liters per kilo of malt). This is a thick mash found only in traditional infusion mashing systems (mash tuns). Such a mass is difficult to stir or agitate and therefore no significant heat-transfer can arise, e.g., to create a temperature program, and it is essential to mix the hot water and cold malt together evenly in a premasher (e.g., a Steeles masher) to achieve an even temperature throughout the mash. Mash tuns are thoroughly preheated before use and heavily insulated to maintain the initial mash temperature constant. Such infusion mashes produce dense worts. The entrained air in the quite coarsely milled malt and air dissolved in the mash water when mixed in a Steeles masher causes the mash to be buoyant (float); this and permits runoff from very deep beds. The thick mash promotes survival of enzymes at the quite high initial heat of an infusion mash (e.g., 65 to 68°C) and of course the enzymes are twice as concentrated at a water:grist ratio of 2.5:1 than they would be at 5:1; the latter is more typical of a lager (or at least, a temperature program) mashing regime.

More dilute mashes are more easily stirred in a mash mixer than dense ones, and so heat transfer is much more efficient; mash mixers usually incorporate large steam coils in their design. A stirred mash can be taken from an initial heat of say 40°C to a mash-off temperature of say 80°C in a short time while following a quite exact temperature profile. Heat transfer from steam jackets is not particularly efficient (though widely used); direct steam injection, direct addition of boiling water or direct addition of boiling (nonmalt) adjunct and returning a decoction (boiled mash) to the malt mash are efficient ways of raising the mash temperature rapidly and accurately. Steam coils can then be used to trim the temperature as required. The hot mash water not used for mashing is used as sparge water in lautering by flowing it through a shallow (9–12 in) grain bed spread evenly over a broad false (perforated) bottom.

Up to this point the solid malt grist and the aqueous wort establish some kind of equilibrium, i.e., material soluble in strong hot wort dissolve and insoluble matter do not (or is precipitated); water–grist contact should not be prolonged beyond this point. In sparging, by contrast, the extracting agent becomes hot water (not hot strong wort) and so it is easy (one might say inevitable) to extract undesirable materials by excessive sparging, especially if the water is unsuitable in composition with, e.g., high alkalinity and low calcium content. When to end wort collection (sparging) is a decision each brewer must make, secure in the knowledge that the extract in last runnings is *not* diluted first worts, it is something else entirely different and contains nothing good. Unused sparge water can be added directly to the kettle to make up the wort volume and adjust gravity, if necessary.

The water drained from spent grain at the end of sparging, and that used in various rinse and transfer processes in the brewhouse, is a polluting load that some brewers prefer to recycle as mash water. This is called "sweetwater" or "weak-wort" recycling.

Wort boiling: Water has a high specific heat and so requires much energy to be heated up (and cooled). Also, it requires about three times more energy to produce steam at 100°C than to produce water at 100°C, because the latent heat of evaporation is required to facilitate the phase change from water to steam. This property also makes steam the most efficient means of transporting heat energy around the brewery: condensation of steam back to water releases much energy).

Maintaining wort at a simmer at say 100°C, rather than at a "full rolling boil", produces unsatisfactory beers. The reason for this is not the limited evaporation of water, although this will affect original gravity (some brewers do boil to a specific gravity specification). Rather, the importance of boiling is the evaporation, with steam, of a host of minor volatile components that arise in wort from the malt, adjunct and hops during brewing processes. The aroma of steam escaping the kettle and its taste, when condensed, testifies to this. Evaporation of wort volatiles therefore requires an energy-consuming, vigorous, steam-generating and water-evaporating boil, and brewers traditionally looked for up 10% evaporation of wort volume. Unfortunately every pound of water evaporated represents a significant input of energy and less evaporation (e.g., 5%), would now be more common. Also, these days much shorter boils are used (e.g., 45 minutes at full boil rather than, say, 180 minutes). A simple way to conserve energy is therefore to minimize evaporation; most obviously, this is done by limiting the excess sparge

Composition of Last Runnings

The temptation for some brewers is to "squeeze' the most extract they can from the brew. They prolong wort separation until the measured specific gravity is extremely low; indeed they may even use the weakest wort to mash-in the next batch of grain.

It is important to realize that the composition of the first and last runnings from a wort separation operation is very different, as the table shows.

Component relative to amount of each component in strongest wort (apart from pH)	Strongest worts	Weakest worts
Fermentable carbohydrate	1.0	0.86
Protein	1.0	1.6
Polyphenols	1.0	1.5
Minerals	1.0	4.0
pH	5.5	6.4

Thus, although the weakest worts still feature sugar as the main source of extract, the less desirable components make a far more significant contribution at the lower gravities.

Uses for Spent Grains

Spent grains emerging from a wort separation system typically have 75 to 80% moisture. Those from a mash filter will be slightly drier than those from a lauter tun. In any case, the cost of drying them is almost invariably prohibitive and the brewer will only do this if there is no immediate opportunity to ship them rapidly away from the brewery: grains present a serious spoilage hazard.

Many suggestions have been made for uses for spent grains, including making into breads and cookies, extrusion into snack foods, growth of microorganisms on them to produce valuable commodities (e.g., xylitol), growth of mushrooms, composting, and fiber board. However, the vast majority of grains still go to cattle feed.

Composition of grains (% dry weight)

Water	8
Fiber	18
Protein	21
Nitrogen-free extract	40
Fat and oil	9
Ash	4

water that must be evaporated. Closed-door boiling also limits evaporation; this departs from traditional practice in which an open kettle door caused a powerful stream of air to cross the wort surface and mix with steam into the exhaust stack; of course the cold air cools and condenses the steam

The Demand for Energy in Brewing

Thermal energy

Brewhouse	45%
Packaging	25%
Utilities	20%
Space heating	10%

Electrical energy

Refrigeration	35%
Packaging	25%
Compressed air	10%
Brewhouse	10%
Lighting	5%
Boiler house	5%
Remainder	10%

and other volatiles leading to inefficiency. In contrast, recent kettle designs seek to promote evaporation of volatiles while minimizing evaporation of water. The two main strategies are (1) to establish means by which all of the wort in a kettle contacts the calandria (heat exchange) surface and boils and (2) to increase the surface area of boiling wort within a steam atmosphere. Under (1) pockets of wort that improperly circulate in kettles (a common design fault) will retain unwanted volatiles and tend to prolong the boil and so increase the amount of evaporation necessary to remove them. Similarly (under 2) if a pump fountains the wort in a kettle that is also boiling, or passes the wort over an inert surface, the opportunity for steam stripping of unwanted volatiles is increased. One major brewery passes boiled wort over a "stripper" with a countercurrent flow of hot air. Energy can also be conserved if it can be usefully recovered from the exhaust stack of the kettle; if this is not done the kettle is a one-way expenditure of energy that is increasingly expensive. The simplest technology is to use exit steam to preheat mash water or to produce hot water for other uses in the brewery. This also condenses the steam and reduces gas/aroma emissions that might be important in some jurisdictions. Vapor recompression has also been tried as a means of energy recovery from the low-pressure steam of the exhaust stack.

Wort temperature must be reduced from near 100°C to cellar (fermentation) temperature quickly. This is mostly done using cold (even refrigerated)

Energy-Saving Alternatives to Wort Boiling

The huge energy load in wort boiling has led to various modifications to this process stage in the interests of making cost savings. These include low-pressure wort boiling (by lowering pressure liquids boil at a lower temperature), high-pressure wort boiling (higher boiling temperatures because of the increased pressure—but for much shorter times) and continuous wort-boiling systems. They have not caught on because of flavor concerns.

It should be remembered that attention to other stages in the process may lead to a reduced need for boiling. These include the use of isomerized extracts and hop aroma essences, mechanical energy (rousers) to replace thermal energy, the use of specific adsorbents for haze precursors and even unwanted flavors and the use of alternative technologies such as high pressure and Ohmic heating.

water or an ice-bank, or other coolant, in a counter-flow heat exchanger. Appropriate choice of plate size and number, flow rates and water temperature permits energy recovery from the wort at about 80% efficiency or so. The recovered water is suitable for mashing and sparging. When cooled, the wort volume shrinks (this must be taken into account in brewhouse calculations), and the specific gravity, viscosity and pH rise somewhat. At the same time the cold break forms.

Counterflow heat exchangers are also (usually) used to cool finished beer to storage/maturation temperature (usually $-2°C$) with the same effects on beer volume, density, viscosity and pH. Beer cooling is part of the refrigeration cycle, requiring energy input for compression of the primary coolant, and is a one-way consumption of energy.

Water again enters beer potentially in two forms in finishing processes: (1) as water for conveying processing aids such as diatomaceous earth or for beer additives or for packing beer transfer lines (to eliminate air see chapter 12) or chase water, and (2) as dilution water for beer made at high gravity. In every case the water must be of brewing quality in all respects, sterile and free of dissolved oxygen. Oxygen is "stripped" from water by creating a concentration gradient down which oxygen can flow. Thus, for deaeration, water can be "sparged" or washed with inert gas or trickle down a tower of inert material with a counterflow of inert gas; water (sometimes saturated with an inert gas) can be sprayed into a vacuum chamber (which is effective in combination with "sparging") or into an atmosphere of inert gas (CO_2 or nitrogen); and steam can be used as a stripping gas, e.g., as in boiling the water, or by spraying hot water into a steam atmosphere. This is expensive but effective and sterilizes the water. In a modern technology water passes across a hydrophobic membrane of hollow fibers with an inert gas on the other side. CO_2 is a useful gas for stripping oxygen because the water for beer dilution must eventually be carbonated (and cold).

Water does not enter beer during pasteurization (heat treatment of beer to achieve sufficient microbial stability), but tunnel pasteurizers use a good deal of water because this is the heating and cooling medium; they are not energy efficient and occupy a large space. In contrast, bulk or flash pasteurization in a heat exchanger with a regeneration stage uses very little water, it heats cellar beer rather than packaged beer and is highly energy efficient and compact. Its disadvantage is that aseptic packaging must follow pasteurization (see Chapter 6). The beer in a package in a tunnel pasteurizer is not agitated, but moves as expected by convection currents, and heat transfer to heat and cool the product, especially through a glass bottle, is relatively inefficient. Packages therefore must spend considerable time in the pasteurizer (e.g., 1 hour or more) so that the heat penetrates to the center-bottom of the package. Containers emerge dry and at the right temperature for labeling. Water and energy conservation is restricted to using, e.g., final cooling water (that is warmed up) to preheat entering cold bottles (etc.), but constant input

Uses for Spent Yeast

Yeast surplus to requirements for repitching has various functions. A proportion is shipped to distillers for their fermentations, where the precise strain is less important than in brewing. A popular treatment in the United Kingdom and Australia is to ship the product for autolysis, the resultant paste being marketed as a spread for domestic consumption. Yeast has long been prized as a vitamin supplement. Yeast has value as an animal feed (the table compares its composition with that of soy meal, another significant animal feedstuff). The yeast must be inactivated prior to consumption, and this may be achieved by addition of propionic acid.

Component (g/100g or μg/g for vitamins)	Brewer's yeast	Soy meal
Protein	36	48
Polysaccharide	30.7	–
Fat	2.6	1.0
Ash	7.3	5.7
Arginine	5.0	7.7
Histidine	3.5	2.4
Isoleucine	5.0	5.4
Leucine	7.5	7.7
Lysine	8.0	6.5
Methionine	2.0	1.4
Cysteine	1.6	1.4
Phenylalanine	4.5	5.1
Tyrosine	4.9	2.7
Threonine	5.0	4.0
Tryptophan	1.0	1.5
Valine	6.0	5.0
Thiamine	200	9
Riboflavin	50	4
Pyridoxine	30	7
Nicotinic acid	500	24
Folic acid	25	4
Pantothenic acid	80	21
Biotin	1	
Vitamin B_{12}		0

Environmental Impacts in the Maltings and Brewery

Impact	Good practice
BOD of waste streams	Mashing using weak worts from previous brew. Mash filter to produce drier grains with higher extract recovery. Mixing precipitates (trub) generated in boiling with spent grains. Yeast collection. Waste beer recovery. On-site effluent treatment.
Heat output and energy consumption	Insulation Maintenance, e.g., steam leaks. Wort boiling procedures (see table below). Heat recovery. Collection and return of condensate in boiler house.
Excessive water usage	Cleaning In-Place optimization. Continuous processes. Recycling of water where appropriate (e.g., coolant water to mash-in; pasteurizer).
Carbon dioxide evolution	Collection off fermenter. Boiler house efficiencies.
Electricity usage	High-efficiency motors. Frequency converters. Energy-efficient lighting. Packaging line efficiency.
Spent filter aid	Centrifuge. Crossflow filtration.

of energy for heating and cooling (as hot and cold water) is required. This water must be treated to prevent "bloom" or water spots on bottles and to prevent growth of microbes in the machine.

WASTEWATER

Maltings and breweries produce a wide spectrum of waste materials that have become increasingly difficult and expensive to dispose of. Primary among these wastes is water because the environment and municipalities have grown more sensitive to it and high prices are charged for disposal. Steep water is intensely polluting and modern strategies that separate barley washing from steeping ameliorate this somewhat. In a brewery the brewhouse produces relatively low volumes of waste water of high polluting load and so do the fermentation cellars, primarily because of the contamination of waste water with wort or beer that can have a biological oxygen demand (BOD) of 100,000 mg/l and suspended solids such grain and yeast. Bottle shops produce a large volume of wastewater

Levels of Effluent

The table gives a guide to the most significant sources of effluent in the brewing process.

	Percentage of total BOD	Percentage of total Suspended Solids
Sweet wort production	7.6	3.4
Kettle	Negligible	Negligible
Hop separators/whirlpool	14.1	11.1
Fermenters	43.7	68.8
Yeast handling	1.4	0.1
Centrifuges	1.5	1.6
Receivers	3.5	0.5
Filters	0.8	8.0
Cold conditioning	12.6	0.8
Packaging	13.8	4.7
Miscellaneous	1.0	1.0

that effectively dilutes the organic load. Throughout malting and brewing processes cleaning and sanitation are part of the effluent problem because such wastes carry a significant organic load as well as the cleaning agents required. In this case, an additional problem is maintaining effluent temperature and pH level within acceptable parameters. In such cases a bulking tank that allows mixing effluents from several sources in the brewery before discharge are useful and waste flow can thereby be regulated; this avoids shock BOD loads and fluctuation of waste volumes at the treatment plant.

The first strategy in control of any waste is to minimize it and/or recycle it in some way, but eventually wastewater must be processed for discharge to the environment or to the municipality for final purification. Brewery wastewater contains primarily organic matter that is rather easily degraded by biological means, eventually to CO_2 and H_2O. Many breweries treat their own waste streams to minimize the costs of disposal that are calculated from volume, BOD load and suspended solids. Anaerobic treatment allows the recovery of about 90% of the carbon in the waste as a flow of a useful gas mixture (methane 70% and CO_2 30%) with considerable BOD reduction, which can be further reduced, if necessary, by traditional aerobic treatment.

Chapter 13

Where the Wastewater Comes From in a Brewery

Place of discharge	Flow of water (liters per hectoliter)
Mashing	3
Lautering	
- last runnings	8
- wash	4
Boiling	
- clean	14
- rinse	3
Whirlpool/ hopback	8
CIP caustic brew	10
Fermentation	
- rinse	8
- clean	2
Surplus yeast	
- if discharged direct to drain	2
- if pressings discharged to drain	1.5
- if sold for food or animal feed	0
Conditioning	
- sediments	3
discharged direct to drain	2.5
pressings discharged to drain	0
pressings to product	10
- washing	
Filtration	
- last runnings	2
- washing	50
Stabilization	
- regeneration	50
Kegging	
- washing	30
- pasteurizer	12
Cask filling	
- returns	1
- washing	< 100
Bottle washing	< 100 up to 3000
Bottle/can filling	< 100 up to 2000
Total *(100% keg production)*	214–219
Total *(80% cask, 20% bottle)*	190–1150

After Rob Reed and Gerry Henderson, *Ferment*, 1999/2000, vol 12(6), 13–17

Costs of Brewing

One set of figures makes a poor approximation to the costs of brewing worldwide: one size does not fit all. For example, the rates of taxation differ enormously globally. Matters of scale are also important: the cost structure for a very small brewing operation will be very different from a multibillion dollar corporation with economies of scale. Equally, their packaging and marketing costs will be far more significant than those for a pub brewer who may be moving beer direct from bright beer tank to pump at the bar. For the latter the raw materials costs will make far bigger proportionate inroads into the budget. The table then is intended as only a very ballpark indication of the relative cost of materials and operations.

Item	Proportion of the cost
Hops	0.2
Adjuncts	1.5
Miscellaneous ingredients (water, filter aids, CIP solutions, etc.)	1.8
Malt	3.5
Production	20
Sales and marketing	20
Packaging	26
Taxation	27

Process Intensification and Economization

A selection of approaches to allow intensification of processes and cost shaving:

Stage	Concept
Mashing	Addition of exogenous β-glucanases and pentosanases to complete the cell wall digestion and release starch
Wort separation	Mash filter to enhance recovery of extract in high concentrations from hammer-milled malt

Stage	Concept
Boiling	Energy conservation by high- or low-pressure wort boiling, mechanical vapor compression, limited evaporation, stripping
Fermentation	HGB (fermentation at high strength with downstream dilution to target alcohol concentration). Continuous fermentation, Foam suppression by ultrasonics to maximize fermenter fills
Maturation	Acetolactate decarboxylase to eliminate diacetyl precursor. Heat to convert acetolactate to diacetyl, followed by passage through immobilized yeast to eliminate the diacetyl produced plus any free diacetyl
Filtration	Crossflow filtration
Stabilization	PVPP, Silica hydrogels, Tannic acid
Finished product	Isomerized hop extracts. Reduced hop extracts to protect against light damage

|14|

Sanitation and Quality

Quality assurance (QA) is a compendium term that broadly encompasses all a brewer's effort to assure a consistent product in the marketplace. It ranges over such diverse things as brewery design, layout and maintenance, brewer education and training, plant sanitation, raw materials quality and controlled processing parameters, quality control (QC) specifications (analytical values), born-on and best-by dates, uncompromising attention to detail, and continual review of quality records and practices, and so on. A more contemporary term might be Total Quality Management (TQM); QA is recognized in such formal programs as ISO 9000/9001, U.K. B55750, U.S. MIL-Q-9858 etc.

It is a fundamental axiom of brewing that consistent high quality of beer arises from raw materials of high and consistent quality, with a consistent process in a protected environment (a brewery) and safeguarded by rigorous sanitation. Brewers build quality partnerships with their suppliers and, once these partnerships are working well they are reluctant to change them. Similarly, the manufacturing process for a particular beer brand is changed rarely and reluctantly and only for good reason and with every safeguard that the flavor will not be noticeably affected. For example, new processing aids or additives or simplified brewing practices usually get their first trials with new products, not established ones. Brewers are indeed consistent and

Chapter 14

Quality Control and Quality Assurance

QC comprises the measurement of parameters that allows a response to those measurements. In other words it is a reactive system.

QA involves designing systems that ensure that a quality product will be obtained through a quality operation. As such, it is a proactive approach that builds robustness into operations such that the chances are strong that on-target performance will be achieved. It is still necessary to perform QC checks, but with a greater assurance that they will merely confirm that specifications have been met. QC is a part of QA, not the other way round. QA embraces issues such as optimized plant design, establishment of standard operating procedures, installation of in-line monitoring and feedback control systems etc. HACCP is a good example of a QA technique.

---

this often translates as conservative when making changes of any kind that might affect the product.

Within the wide range of topics included in QA two relate to the theme of this book—sanitation and QC—and these apply at every step of the process from barley harvesting to beer packaging.

QUALITY CONTROL

The many aspects of QA or TQM as outlined in the opening paragraph vary widely among different companies; however all companies rely on stringent QC procedures as a central tenet within QA. QC is a series of numerical standards, or specifications, for raw materials, process parameters and product qualities that can be reliably measured and used to determine whether or not the process remains "in control" over time. Brewers usually choose methods tested and recommended by their national brewing authorities, e.g., ASBC in the USA; IBD in the UK; Analytica-EBC in Europe; MEBAK in Germany. Though some are "international" methods most of them are not identical or interchangeable. However, many analytical methods gain acceptance in practice long before these agencies

Standard Methods

In order that different laboratories (e.g., different labs within one company; labs of suppliers and customers) can strive toward finding agreement on analyses, standardized methods have been established by several organizations. The main ones are those from the American Society of Brewing Chemists, the European Brewery Convention and the Institute of Brewing and Distilling.

Relevant methods are debated in committee and written in a standardized format that is clearly intelligible, allowing for the least possible error in pursuit wherever it is operated. The method and samples for measurement are circulated to a wide range of laboratories, which individually produce a set of data. This is collated and analyzed statistically centrally, prior to the assigning of values for repeatability (r_{95}) and reproducibility (R_{95}).

'Little r' (as it is called) is a measure of how consistent the results are when a method is applied by the same analyst in a single location. It is defined as "the difference between two single results found on identical test material by one operator using the same apparatus within the shortest feasible time interval will exceed r on average not more than once in 20 cases in the normal and correct operation of the method" and is given by $2 \times \sqrt{2} \times \sqrt{(\sigma_r)}$, where σ_r is the SD for the procedure when assessed within a single laboratory.

"Big R" is an index of how good the agreement is when a method is applied to the same sample but in different laboratories with different analysts. It is given by $2 \times \sqrt{2} \times \sqrt{(\sigma_b^2 + \sigma_r^2)}$, where σ_b is the between lab SD.

Only if these values are acceptably low will any confidence be placed in a method for its ability to give reliable and reproducible values that can be used not only for process control but also as a basis for transactions. If values for r_{95} and R_{95} are good, then the method will be added to the recommended list of methods.

approve of them mainly because the approval process is long and quite arduous. Some methods that brewers commonly use are unlikely ever to be approved by these agencies, because brewers adopt some "in house" methods that work well only in their particular circumstance. In all cases, brewers strive to use methods that give reliable, relevant, useful and actionable information.

Quality control methods should be simple to conduct, accurate, repeatable and reproducible and inexpensive. Brewers and suppliers work hard to assure that methods yield consistent numbers, on the same samples, from time to time (repeatable or precise) and from place to place (reproducible). Methods must be not only precise but also accurate, that is, reflecting the "true" value of the sample, and most are accurate. Nevertheless methods that are reproducible and repeatable (precise, i.e., having consistent bias) can still be useful though inaccurate. Many analyses brewers use fall into this category.

Precision and Accuracy

Liken an analytical method to the game of darts. The first thrower groups all her darts together on the dartboard, but the close grouping is some distance from the target (bull's-eye). Clearly she is a very precise but not accurate. The second player, aiming for bull's-eye, lands the darts equidistant from the bull but arranged right the way round the board and with no degree of consistency. This person is accurate, insofar as on average he is closer to the true mark than is the other thrower, but he is much less precise: there is no predictability about where the next dart will land. The ideal is to hit the bull's eye every time, but of the scenarios described above the more desirable is precision. If an analytical method properly followed gives values that are reliable and meaningful, then it is very much of secondary significance whether they are the "true" values. Provided we have defined terms of reference (standards) against which we can compare our data, then we can achieve control using such a procedure.

An example would be the measurement of protein in grain. For many years this was quantified by the Kjeldahl procedure, but concerns about health and safety led to its replacement by the Dumas method. The latter, which involves total combustion of protein, leads to higher values for protein for any batch of grain than does the Kjeldahl method. The cereal is not changed; it is merely the method that has changed. It was necessary for the coordinates to be established slightly differently, and for brewers and maltsters to think slightly higher when it comes to protein.

Sanitation and Quality

No method is perfect (absolutely precise) and all methods yield a range of values even when repeated on a single sample; this variation is called error. In general, error arises from three sources: sampling error + systematic error (bias) + replicative error. The first arises from the samples; the second arises from the method itself, the instruments used and the actions of the analyst(s); both of these sources of error can be minimized by appropriate action. Replicative error occurs randomly and without explanation.

Proper sampling is a key to successful QC operations; this is more difficult than it sounds, especially for particulate materials such as malt, hops and perhaps yeast. A sample must be sufficiently large (or frequent), representative of the whole lot, and taken randomly, and on this depends the usefulness of the results obtained. Similarly, analysts must be well equipped, appropriately trained and properly supervised because adherence to every detail of standard methods is crucial to their efficacy. Under these circumstances, calculating the arithmetic average or mean (x-bar) of multiple measures (more is better) on each sample probably gives a good measure of the "true" value of an analyte. We would expect these numbers to vary randomly (i.e., by chance alone = the replicative error) about the mean, with small variations being more common than larger ones; this is called a normal distribution that, when plotted out, gives a bell-shaped curve that may be more or less broad (Figure 14.1). The standard deviation (SD) is the best estimate of the random error in a method and is directly related to the spread of the distribution; a small SD of a method (narrow spread) is preferred to a broad one. SD is often expressed as the coefficient of variation (CV) of the method and expressed as a percentage.

In any normal distribution we expect virtually all results (99.7% of them) to lie within ±3 SD of the mean, 95.4% within ±2 SD of the mean and 68.35% within ±1 SD of the mean. Thus we can expect that, for any method

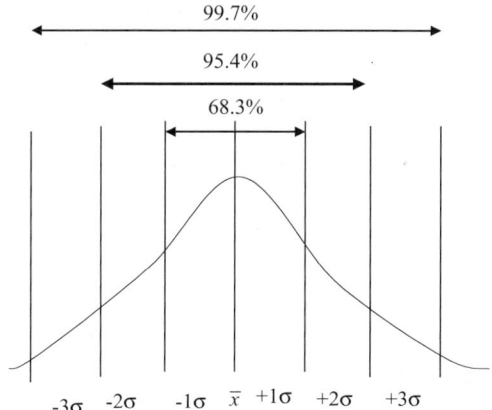

Figure 14.1. standard distribution.

by chance alone, a single value will deviate more than 1sd from the mean only once in about three measures (100/100−68.35) and 2 SD from the mean only once in 21 measures (100/100−95.4). This can happen twice in succession (for 1 SD) about once in 10 measures and for 2 SD about once in 440 measures. Deviations greater than this, therefore, suggest that factors other than random error are involved. Further, note that SD is for individual measures; if multiple measure are taken and the SD of those is calculated it is called the standard error of means (SEM) and is much smaller than SD (but note that it applies to means of numerous measures, not to single measures). Thus, the SEM for, e.g., four measures made on a single sample is half the SD for a single measure on the same sample. This means that analysts are much better off using multiple measures and taking the mean than relying on single measures. This relates to a central tenet of statistics called the Central Limit Theorem and underpins the use of QC charts. Charts may be based on deviation from the mean (process control mean chart, the most common) or from the acceptable range of data.

QC charts are only useful if well designed, well maintained and properly interpreted and used; reading them for action usually follows the application of three simple "rules": action must be taken (1) if a single data point lies outside the chosen action limits (e.g., 3 SD or 3 SEM); (2) if two out of three successive points lie outside the warning limit (e.g., 2 SD/2 SEM) or (3) if there is a run of eight data points on one side of the mean. Conversely, a process is considered "in control" if (1) no value lies outside the action limits, (2) no more than about one value in 40 is outside the warning limits, (3) there is no instance of two consecutive values lying outside a warning limit, (4) there is no sequence of five or more values that infringe a warning limit, (5) no more than six values lie on the same side of the mean, (6) there are no runs of more than six values that are rising or falling (trends).

Charts such as these are useful for presenting the results of methods that monitor continuous variables, i.e., specific qualities of the process or product over a period of time such as pH, color, haze, starting and end gravity, bacterial counts, oxygen content etc. Other QC methods measure the qualities that a material or product has, e.g., flavor of beer, barley protein, diastatic power in malt or α-acids in hops etc. These data are recorded in tables that can be matched against specifications. Typical analyses for barley, malt, hops, water and beer are given in table.

SANITATION

While it is probably quite possible (even likely, most of the time) to produce a clean food product in an unclean plant, this is neither desirable nor legal. The FDA defines adulterated food (with other definitions) as food

Some Basic Statistical Concepts

The *mean* value for a set of data is the average of the values, obtained by adding them up and dividing by the number of measurements.

If the individual measurements are x_1, x_2, x_3 etc., the total number of measurements is n and the mean is M, then

$$M = \frac{1}{n}\sum x_i$$

The *range* is the difference between the highest and lowest values. This tells us nothing about how the values are distributed within that range as we can see if we plot the individual data in a histogram or a curve.

For any individual measurement the extent to which it deviates from the mean is given by the expression $x_i - M$. Summing all the deviations for an evenly distributed set of data will result in a value of 0 for there are as many values higher than the mean as there are below it. The problem is eliminated by squaring the $x_i - M$ value and incorporating it into the SD (σ):

$$\sigma = \sqrt{\frac{1}{n-1}\sum (x_i - M)^2}$$

The lower the SD, the more precise is an assay. The coefficient of variation (CV) is a simple way to illustrate this variation:

$$CV = \frac{100\sigma}{M}$$

It allows expression on a percentage basis of the error inherent in a method. The lower is CV, the more reliable is the analysis.

Usually, for a sufficiently large data set, data adopts a "normal distribution." There is a 68.3% chance of the value being within 1 SD of the mean, 95.4% probability of it being within 2 SDs and a 99.7% chance of it being within 3 SD.

$$\text{Process capability} = \frac{\text{upper limit of measurement} - \text{lower limit of measurement}}{6\sigma}$$

> Obviously if the difference between the upper limit and lower limit is small then we have a very narrow data spread.
>
> Brewers frequently have warning and rejection limits for individual parameters. The warning lines are set on the basis of standard errors of measurement (SEM), where
>
> $$SEM = \frac{\sigma}{\sqrt{n}}$$
>
> Warning values are at two standard errors ($2\sigma/\sqrt{n}$) above and below the target. The reject lines are set at three standard errors ($3\sigma/\sqrt{n}$) above and below the target.

produced under conditions "whereby it *may have* become contaminated"; this means that plant inspection as well as food analysis is a proper means of monitoring sanitary operations. There is an apocryphal story of a brewer who inspected his suppliers' premises armed with a loaf of sliced bread; he felt free to wipe any product-contact surface with bread and demand that the supplier eat it! Such disrespect of brewer for supplier has long gone (if it ever existed!), but the story serves well to illustrate the interdependence of brewer and supplier, and, throughout all parts of the process of beer making, the need for pristine operations that are regularly and critically inspected.

The word sanitation derives from the Latin *sanitas* that implies an healthful and wholesome foundation to food-processing operations, including maltings, hop yards and breweries. In this broader sense of "good housekeeping" sanitation affects the environs of the plant and process including, e.g., the plant layout, construction and ship-shape organization and general cleanliness (free from dust and spillage, for example), control of rodents, insects and birds especially around grain-handling operations, as well as the process itself; this last factor concerns mainly the product-contact surfaces that, before use, need to be free of all soil and microbes and concerns us here.

Sanitation is necessary at every stage of the process. The physics of soil removal imply that an absolutely clean surface is impossible to attain, and so, in practice, sanitation is about making beer contact surfaces *sufficiently* clean. This means that different practices suffice depending on the location in the maltings or brewery. It is therefore best to make some general observations that apply to every stage, recognizing that quite different sanitary strategies might be necessary from place to place. Sanitation is a planned preventative program; such a plan is sometimes referred to as the brewer's lifeline. Sanitation is not a trouble-shooting program because, if it is, it becomes part of the problem. Thus, problems are most easily resolved from

HACCP

HACCP (Hazard Analysis and Critical Control Points) is a QA system focused on improving the microbiological robustness of an operation such as brewing. Exactly analogous systems can be instituted for any other facet of quality and process control, not just microbiological ones.

Stage one is to develop a detailed process flow diagram and then to analyze each stage for the microbiological risk it presents. These risks are then ranked in order of their threat and the critical control points (CCP) are identified. A CCP is a stage that has a significantly deleterious impact on the product. It is necessary then to set the tolerance limits for each of these CCPs and to establish procedures for monitoring at the CCPs. Examples might include the monitoring of CIP detergent strength and pasteurizer times and temperatures. Fifth is the setting of corrective action protocols. Finally procedures need to be set in place for verifying that the system is functioning and for maintaining records and documentation.

---

a basis that knows that sufficient sanitary procedures were in place and scrupulously followed. Thus a sanitation plan that is reasonably detailed should be in place for all areas of the brewery and/or for individual pieces of equipment or for processing events, and should include (1) the objective of cleaning/sanitizing, (2) the method(s) to be used, (3) its frequency of application, (4) any dismantling required etc., (5) how the results should be monitored and (6) what comprises satisfactory sanitation/cleaning in each case. Plans will differ quite considerably for cleaning a wort kettle, for example, where the prime concern is removing soil from the surfaces that might impeded heat transfer in the calandria, compared to a Bright Beer Tank (BBT) where the "soil" is merely cold beer. The following paragraph will focus briefly on cleaning and sanitation of beer contact surfaces.

Cleaning and sanitation are different technologies with different objectives, and should neither be confused with each other nor one substituted for the other. In all cases cleaning should precede sanitation because cleaning removes soil (and with it most of the microbial load) and so promotes the efficacy of the sanitizing action that follows, i.e., adequate sanitation is only possible if applied to a clean surface. Cleaning is best done with soft water (or with cleaners formulated to sequester calcium) as this avoids the deposition of "beer stone" on surfaces. Cleaning can vary depending on the amount and nature of the soil involved from aggressive (e.g., hot caustic solutions in kettles) to mild (e.g., an occasional acid rinse of BBTs). But in any case the sequence of events is the same: pre-clean rinse, clean, post-clean rinse, sanitize. Generally, brewers clean a tank immediately after use and sanitize it immediately before use. Cleaning involves an input of energy required to break the bond between soil and surface and can be in three forms: heat energy, mechanical energy and chemical energy applied over a sufficient period of time. These factors can substitute for each other so that a brewer can operate a cleaning cycle that fits in with the tempo of the brewery and/or manages cleaning costs wisely. Cleaning and sanitizing these days is mostly done by CIP (cleaning in place) systems in which the brewery is hooked up in cleaning loops to vessels that store and maintain cleaning solutions. When properly managed such systems increase the reliability of cleaning and reduce costs and the impact of cleaning on the environment by reducing chemicals discharge and by recycling rinse waters.

Index

β-glucan, 4, 34, 43, 56, 96
β-glucanase, 100, 106, 112
   assay for, 107
α-di-carbonyls, 137
α-helix, 6
α-oxo-acids
   decarboxylation of, 122
   detoxification of, 122
β-pleated sheet, 6
2,6-dichlorophenolindophenol, 140
2-trans-non-ene-al, 135
4-vinylguaiacol (4-VG), 64

a * value, 26. See also CIE-L*a*b* method
Abmeter, 117. See also Coulter particle counter
acetoin butanediol, 62
acetyl-coenzyme-A, 63, 122, 125
Achilles heel, 109
acyl-CoA, 122, 133
adjuncts, 89
   malts and, 90
   yellow corn grits, 89

adulterated food, 166
alcohol content
   and freezing, 49
alcohol dehydrogenase
   role of zinc for action of, 71
alcohol-chilling test, 53
aldol condensation, 138
ale fermentations
   role of Obesumbacterium in, 65
ale yeast, 115
aleurone layer, 7
algebraic notation of pH, 13. See also pH
alkalizing effect, 16. See also phosphoric acid dissociation
all-malt beer, 30
Amadori rearrangement, 24
America Society of Brewing Chemists, 37, 163
amino acid transaminase reactions, 102
amino acids, 4
   buffering power of, 12
   ionizable groups of, 15
   sequence of, 6
aminopeptidase, 7. See also exoenzymes
amphipathic polypeptides, 37

amylases
  and starch-digestion, 86
amylo-glucosidase, 90
anaerobiosis, 59, 131
antioxidants, 21, 138
arabinoxylans, 99
ASBC. *See* America Society of Brewing Chemists
aspartate buffer, 17
ATP bioluminescence test, 66
autolysis, 9

b * value, 26. *See also* CIE-L*a*b* method
barley
  absorbance spectrum of, 26
  active forms of oxygen from enzymes for, 136
  and endosperm cells, 3, 5
  and rule-of-thumb as 13/13, 80
  and use of gibberellic acid, 82
  general kinds of, 6
  genetic modification of, 89
  malting of
    cell wall degrading enzymes from, 96
    lipid-degrading enzymes from, 135
    protein-degrading enzymes from, 8
    starch-degrading enzymes from, 101
  modification measurement methods, 93
  moisture content of, 81, 143, 145
  polyphenol oxidase, 134
  protein levels in, 5
  steeping, 82, 144
    objective of, 144
barley identification
  gel electrophoresis for, 6
Bavarian weizenbiers, 64
BBT. *See* Bright Beer Tank
beading, 33, 35. *See also* creaming
beer
  4-vinylguaiacol in, 64
  and bubble skins, 30. *See also* skeins
  and net foam quality, 29
  and resistance to spoilage, 68
  and silica, 73
  and source of nutrients, 99
  bits in, 45
  color, 20–27
    addition of caramels for, 25
    CIE-L*a*b* measurement method, 25
    color space, 25
    flavor, 20
    kettle boil for, 23
    Maillard reaction, 20

beer (*cont.*)
  color (*cont.*)
    melanoidin pigments, 20
    polyphenol oxidation, 20
    quality criteria, 20
    spectrophotometry, 25
  contact surfaces, 169
  contamination of, 59
  different polypeptides present in, 29
  Enterobacteriaceae related to processing of, 72
  flavor stability prediction of, 139
  foam dispersed phase, 30
  foam problems of, 40
  foam, 28
  grist
    adjuncts as, 24, 56, 89
    malts as, 24, 87, 90
  haze, 6, 12, 41
  infection of, 59
  making
    role of inorganic ions in, 76
    role of zinc in, 71
    transformations in, 105
  microbial contamination of, 58
  protein concentration of, 31
  reducing power of, 137
  skeins, 30
  spoiling bacterias
    gram-negative/positve bacteria, 61, 72
    heterofermentative bacteria as, 61
  stability, 160
Big R, 163
biological catalysts. *See* enzymes
biological oxygen demand (BOD), 156, 157
Blom principle, 38
boiling of wort, 48
Bradford method, 5
brewery
  brewhouse, oxygen from elimination of, 133
  cleaning and sanitizing of, 170
  fermentations of, 116
    and contamination, 63
  oxygen pick-up, 46
  source of wastewater in, 158
  water and energy connected to, 143
brewing process, 3
  barley in, 3
  breaks
    cold/hot break, 48
  commercial enzymes used in, 112
  compendia, 60

brewing process (*cont.*)
   conditions and microbes, 59
   costs of, 159
   denaturation of proteins, 6
   effluent in, 157
   energy demand for, 152
   enzyme activity in, 109
   enzyme conversion, 6
   enzyme–substrate system in, 6
   gene technology, 113
   haze-forming materials in malt as, 44
   hops, 77
   malt kilning, 6, 124
   mash temperature program, 6
   measurement of N-containing materials in, 4
   oxygenation in, 132
   raw materials for, 77–92
   role of adjuncts in, 89
   steeping in, 143
   temperature control in, 109
   waters, 16
      and calcium content, 16
      and mash pH, 17
      pH active principle of, 16
   wort boiling, 124
   yeasts
      taxonomic categorization of, 114
bright beer tank, 169
brink, 9
BU value, 80
buffering agents, 17
building block, 4
Burton salts, 70, 72, 124

calandria, 153
Calcofluor, 95
caramel formation, 24
carbon dioxide
   in beer and creaming, 33
carboxypeptidase, 7. *See also* exoenzymes
cardboard, beer flavor, 136. *See* papery
carrageenan, 9. *See also* kettle finings
celiacs, 56
cell wall model, 98
cellar wort, 17
centipoise, 97
Central Limit Theorem, 166
chelation, 71
chemical catalysts, 105. *See also* enzymes
chemiluminescence
   measurement of, 140
chill-stable beer, 5

CIE-L*a*b* method, 26
   absorbance spectrum of beer in, 26
cold stabilization, 49
cold storage, 48
cold-water extract (CWE)
   and protein breakdown products, 5
colorimetric methods, 5
column chromatography, 4
commercial enzymes, 112
   β-glucanase, 112
   amylases, 112
   glucoamylase, 112
comparator methods, 25
   lovibond tintometer in, 25
competitive inhibition, 111
condiment brewing, 27
contamination
   microbes for beer
      fortune conditions for, 59
      regulating guidelines for, 148
conversion temperature, 8
cooled wort
   and cold break, 48
Coomasie Brilliant Blue, 4, 38
Coulter particle counter, 117
cP. *See* centipoise
creaming, 33
critical control points (CCP), 169
CV. *See* coefficient of variation

Darcy's law, 97, 99
DCPIP. *See* 2,6-dichlorophenolindophenol
De Vries equation, 33. *See also* disproportionation
dead bacteria, 46. *See also* haze
decarboxylation
   ferulic acid of, 64
decoction mashing, 104
DEFT. *See* direct epifluorescent filter technique
DeMan Rogosa Sharpe, 61
denaturation, 6, 109
deoxynivalenol, 41
diacetyl pathways, 62
   and production, 62
diastatic power, 86, 107
dimethyl Sulfide (DMS)
   control of, 103
dimethyl sulfoxide (DMSO), 102
dipeptide permutations, 4
direct epifluorescent filter technique, 66

disproportionation, 32, 33, 34
　and De Vries equation, 33
　film thickness in, 33
divalent ions, 16
divalent metal cations, 37
DMS levels
　factors impacting, 126
DMS precursor, 87, 126
DMSO reduction, 104
DON. *See* deoxynivalenol
DP. *See* diastatic power
Dumas method, 3, 4, 164

electron spin resonance technology, 140
EMP pathway, 118
endogenous antioxidant (EA) value, 140
endogenous inhibitors, 111
endopeptidase enzymes, 7
　sulfhydryl enzymes, 7
endosperm
　and aleurone layer, 7
　mealy, 144
energy in brewing
　electrical energy, 152
　thermal energy, 152
Enterobacteriaceae, 65
enzymatic browning, 23. *See also* polyphenol oxidation
enzymes, 105–113
　and kinds of reactions, 110
　and pH environment, 108
　and rate of the reactions, 106
　assays, 107
　　and diastatic power, 107
　conservation, 6
　denaturation of, 109
　factors responsible for, 109
　inhibition types of, 111
　proteins, 108
　reaction catalyzed by
　　product formation in, 109
　reaction elements of, 105
　temperature optima for, 110
enzyme–substrate system, 7
equilibrium constant K, 14
ESR. *See* electron spin resonance technology
ethyl acetate, 122
European Brewery Convention, 163
exoenzymes, 7
　aminopeptidase, 7
　carboxypeptidase, 7
extract yield
　factors impacting, 129.

FAN. *See* free amino nitrogen
FCT. *See* foam collapse time
FDA, 166
fermentation, 115
　application of amylo-glucosidase in, 90
　factors impacting, 129
　inputs and outputs in, 116
　practices of, 128
　rate of, 120
　temperature for, 120
ferulic acid
　and 4-vinylguaiacol, 64
flash pasteurization, 154
flavoring compounds
　and flavor stability
　　factors impacting on, 141
　factors impacting, 125
flint bottles
　beer in, 20
foam, 28–42
　and emulsion of gas, 30
　assessment methods of, 37
　bubble formation, 32
　　and radius of nucleation site, 32
　　and relative density of the beer, 32
　　bubble radius in, 32
　collapse
　　and Nibem meter, 38
　　mechanisms for, 30
　impacts, 39
　model, 37
　　foam collapse time in, 37
　　hydrophobic interactions in, 37
　　normalized half-life in, 37
　　sigma (Σ) value, 38
　nucleation site in, 32
　physical events in, 32
　proteins, 28
　quality of
　　proteolytic enzymes for lessen, 55
　stability, 28, 34
　　beer viscosity, 34
　　divalent metal cations, 37
　　haze potential, 43
　　model for, 37
　　nitrogen gas for, 37
　　polypeptide content of beer, 28
　　propylene glycol alginate, 37
　stabilizers used for
　　albumin-derived polypeptides, 55
　　hordein-derived proteins, 55
　volume
　　conductivity to estimate, 38

Index

foam–beer interface, 38
foam-negative agents
  detergents as, 31
  lipids as, 31
foam-positive proteins, 10
foamstabilizing complexes, 30
Folin–Ciocalteau method, 5
food-processing operations, 168
free amino nitrogen, 5
  content of, 8
  values, 100
friabilimeter, 94
full rolling kettle boil, 79
Fusarium
  deoxynivalenol produced by, 41
  hydrophobin produced by, 41

GA3. *See* gibberellic acid
gallotannin, 55
gel electrophoresis, 6
germinating malt, 7
  peptidolytic enzymes in, 7
  proteolytic enzymes in, 7
gibberellic acid, 82
glutamate buffer, 17
gluten free beer, 56
glycolysis, 111, 118. *See also* EMP pathway
gram-negative organisms, 60, 65
  Acetobacter spp, 65
  Gluconobacter spp, 65
  Megasphera spp, 65
  Pectinatus spp, 65
  Zymomonas spp, 65
gram-positive organisms, 60
grist color, 22
growth media, 60
  differential media, 60
  general media, 60
gushing, 41
  and nucleation sites, 41
  common cause of, 41

hazard analysis and critical control points (HACCP), 169
haze, 43–57
  active proteins, 53
  analysis of
    identification of, 44
      pitfalls of identification and staining techniques for, 44
  and generic polyphenol, 52
  and generic protein, 52
  common cause of, 43

haze (*cont.*)
  in beer, 46
  measurement, light scattering method for, 51
  potentiating
    polyphenols, 44
    proteins, 44
  precursors used, 56
  preventative for
    papain used as, 55
  protein that form
    silica hydrogel for removing, 54
    xerogels for removing, 54
  proteins, 10
  shelf-life
    and forcing methods, 52
  stability of beer, 28
    precipitation tests for measuring, 53
    prediction of, 53
  standards of, 51
    role of hexamethylenetetramine in, 51
    role of hydrazine sulfate in, 51
head retention value, 37
hemocytometer, 117, 120
Henderson–Hassebalch equation, 15
heterofermentative bacteria, 61
heterogeneity, 4
hexamethylenetetramine, 51
HGB. *See* highgravity brewing
high performance liquid chromatography, 41
high-adjunct beer, 30
higher alcohols formation
  deamination for, 8
  decarboxylation for, 8
highgravity brewing, 120
hop
  α-acids of, 79
  acids, 30
  and beer stabilization, 78
  assessing resins in, 78
  bitterness to beer, 78
  conversion of compounds on boiling, 30
  electrical conductivity in, 78
  kettle efficiency, 79
  products, 78, 81
  ratio of A275 to A325, 78
  resins, 136
  spectrophotometric measurement of, 78
  storage index, 78
  utilization, 79
hopping rates, 79
hot break, 47
hot trub, 9. *See also* mashing

hot/cold cycling, 53
HPLC. *See* high performance liquid chromatography
HRV. *See* head retention value
hydrazine sulfate, 51
hydrophobic protein particle, 3
hydrophobin. *See* hydrophobic protein particle

imino acid, 6
incineration in oxygen, 3
inorganic ions, 69–76
  and speciation, 71
  calcium, 69
  phosphate, 69
Institute of Brewing and Distilling, 163
inter alia effect, 141
intramolecular bonds, 6
invisible haze, 51
  causes for, 51
isinglass, 10
  and functions, 10
  and protein preparation, 10
  dried swim bladders for, 10
  finings, 10, 45
    and flocculation of yeast, 10
  lipid-binding capability of, 10
iso-α-acids, 36
  antimicrobial properties of, 79
isoelectric point, 9

kettle efficiency, 79
kettle finings, 9
kilning, 6, 145
  and evaporation of water, 145
  and Maillard reaction, 145
  early-drying stage in, 22
  purpose of, 145
Kjeldahl method, 3, 164
KMS. *See* potassium meta-bisulfite
Kolbach index, 93. *See also* soluble nitrogen ratio
kraüsen fermentation, 124
kraüsening, 121

L* value, 26. *See also* CIE-L*a*b* method
Lactobacillus
  thermophillic strain of, 17
lagering, 121
lauter run-off, 8
lautering, 8, 150
lipid transfer proteins, 29

lipids reaction
  with oxygen, 135
lipoxygenase (LOX), 135
little r, 163
lovibond tintometer, 25. *See also* comparator methods

Maillard reaction, 25, 87, 141
  and Amadori rearrangement, 24, 25
  and higher pH, 21
  nonoxidative nature of, 23
  oxidative reactions in, 21
  pH-changing conditions for, 23
  precursors for, 21
  products of, 21
  requirements for, 24
  role of reducing sugars in, 21
  role played by pentose sugars in, 24
malt
  and grist color, 22
  and non-N adjunct, 31
  cell wall modification assessment of, 95
  color of
    and higher flavor, 21. *See also* beer
    and low enzyme content, 21
  FAN, 8
  flavor descriptors for, 86
  friability, 94
    and friabilimeter, 94
  gibberellin treatment for, 22
  phytases in, 70
  protein, 8
  specification
    components of, 87
malt/adjunct ratio, 124
malting, 5, 86
  and breweries
    environmental impacts in, 156
    waste materials produced by, 156
  and low nitrogen content of barley, 5, 55
  enzyme action in, 83
  kilning in, 145
  nitrogen content in barley, 5
mashing
  and cold break, 9
  and use of *Lactobacillus*, 17
  auto-oxidation during, 136
  downstream processes in, 136
  hot break formation, 9
  malt lipase action, 134
  oxidative reactions in, 47
  regimes, 5

mashing (cont.)
  staling aldehydes formed in, 136
  system
    mash-off temperature, 149
    oxygen into, 46
    sweetwater recycling, 149
    temperature program, 8
    trub formation, 9
mash-off temperature, 149
melanoidin pigments, 21, 24
  imidazole rings in, 21
  pyrazine in, 21
  reaction catalyzed by, 138
metabolic flux, 118, 120
metabolic intermediates, 121
methionine sulfoxide (MetSO), 104
microbial spores, 58
microbiological methodology
  ATP bioluminescence test, 66
  direct epifluorescent filter technique, 66
microbiology, 58
  resistance to spoilage of beer, 68
monomeric units, 4. See also amino acids
MRS. See DeMan Rogosa Sharpe
myo-inositol, 102

N-containing materials, 4
  measurement of
NHL. See normalized half-life
Nibem meter, 38
ninhydrin, 4
nitrosamines
  nitrosodimethylamine, 103
noncompetitive inhibition, 111
nonenzymic browning, 87. See also Maillard reaction
normal malting, 5
normalized half-life, 38

Ostwald ripening. See disproportionation
oxygen, 131–142
  affect on beer quality, 131
  measurement of, 132
oxygenated yeast
  and pitching, 133

papery. See cardboard
Pascal-seconds, 97. See also centipoise
pasteurization, 55, 67, 154
  high-temperature-short-time strategy for, 67
Pediococcus spp, 61. See also beer

peptides
  and bonds, 4
  buffering power of, 12
peptidolytic enzymes, 7
PGA foam stabilizer, 45
PGA. See Propylene glycol alginate
pH (hydrogen ion concentration), 13–19
  algebraic notation of, 13
  and buffering reactions, 16
  conjugate acid–base pair for, 14
  exponential scale of, 13
  impacts, 18
  in brewing
    and interaction of inorganic ions, 70
  measurement, 14
    and equilibrium constant K, 14
    and Henderson–Hasselbalch equation, 14
  meter
    and measuring electrode, 14
  pH meter for, 14
phenolic acids
  benzoic acid series, 137
  cinnamic acid series, 137
phlobaphenes, 137
phosphatase enzymes, 70
phosphoric acid dissociation, 15
  and alkalizing effect of bicarbonate, 15
pH-reducing strategy
  use of *Lactobacillus* for, 17
phyate
  dephosphorylation of, 102
pitching yeast, 58, 133
pK value, 101
polypeptides, 4
  albumin derived, 29
  hordein derived, 29
  hydrophobic nature of, 37
  tripeptides, 4
polyphenol oxidation, 9
  peroxidase role in, 23
  polyphenol oxidase role in, 23
polyvinylpolypyrrolidone, 52
potassium meta-bisulfite, 23
precipitation tests, 53
  SASPL test, 53
prefermenter, 9. See also brink
process
  flow diagram, 169
  intensification, 159
product inhibition, 111
prolyl endopeptidase, 55

propylene glycol alginate, 38
protein assessment
  methods involved, 5
  and Dumas method, 3
  and Kjeldahl method, 3
protein Z, 29
proteinolysis, 11
protein–polyphenol
  haze, 48
    and cold storage, 48
    characteristics of, 49
  interactions
    model for, 50
proteins, 3–12
  barley and malt in
    evaluation of, 4
  biological function of, 6
  breakdown, 5
  concentration and foam stability, 31
  degrading enzymes, 8
  foam-positive proteins, 10
  haze proteins, 10
  impacts, 11
  in beer, 5
  precipitation, 8
    affect of pH in, 9
    and Coomasie Brilliant Blue, 4
    and dropping system, 9
    and isoelectric point, 9
    and kettle finings, 9
    and protease release, 9
  roles played in beer, 6
  solubility of, 9
  three-dimensional structure of, 6
  total combustion of, 4
proteolytic enzymes, 7
pseudo haze. See invisible haze
PVP. See polyvinylpyrrolidone monomer
PVPP. See polyvinylpolypyrrolidone
pyruvate decarboxylase, 102
pyruvate, 62

quality assurance (QA) system, 161
  and HACCP, 169
  critical control points, 169
  ISO 9000/9001, 161
  U.K. B55750, 161
  U.S. MIL-Q-9858, 161
quality control (QC) charts, 166
quality control (QC) methods
  arithmetic average of multiple measures, 165

quality control (QC) methods (cont.)
  Central Limit Theorem, 166
  errors, 165
quality control (QC) procedures, 162. See also QC methods
quality control (QC), 161
  and QA, 162
  methods for, 164
  statistical concepts in
    basic, 167
    mean value, 167
    normal distribution, 167
    range, 167
    standard deviation, 165
    standard distribution, 165
    warning values, 167

rancidity, 135. See also lipids reaction
rate-limiting reaction, 121
raw materials, 77
redox potential
  measurement of, 141
reductones, 24, 25, 137
  ascorbic acid as, 137
regression equations, 5
repeatability value (r95), 163
replicative error, 165
reproducibility value (R95), 163
rule-of-thumb as 13/13, 81

salt-insoluble protein. See storage proteins
sampling error, 165
sanitation
  and food-processing operations, 168
  and QC, 162
saturated ammonium sulfate precipitation limit (SASPL), 52
scavenging corks
  and metal-catalyzed oxidation, 141
SD. See standard deviation
SEM. See error of means
shelf-life of beer, 52, 54
  strategy to increase, 52
siebert model, 50
  and protein–polyphenol interactions, 50
  for haze formation, 50
sigma (Σ) value, 38
soluble nitrogen ratio (SNR), 93
soluble proteins, 6
soluble surfactant, 30
spectrophotometer, 4
spent grains, 151

Index

spent yeast, 155
spread-plating, 65
–S–S–bridges, 6
stabilizers, 54
staling
   aldehydes, 136
   substances pathways to, 138
starch, 4
sterile filtration, 67
storage proteins, 6
Strecker degradation, 24, 100, 137
substrate and enzyme
   binding between, 108
substrate inhibition, 111
sulfhydryl enzymes, 7
superoxide dismutase, 136
   role of zinc for action of, 71
sweet wort
   production of, 11
sweetwater recycling, 150
systematic error, 165

TBA. *See* Thiobarbituric acid
temperature optima, 110
thermolysis, 24
thiobarbituric acid, 140
three-dimensional structure of proteins
   and intramolecular bonds, 6
total quality management (TQM), 161
tripeptides, 4. *See also* polypeptides
trub flocs, 9

VDK precursors, 121
viability test for yeast, 120
   and hemocytometer, 120
vicinal diketones (VDK) levels, 121

water and energy, 143–160
   conservation of, 143
water evaporation
   latent heat of, 145
water regulations, 146
weak acids
   and ability to buffer, 16

weak bases
   and ability to buffer, 16
weak-wort recycling, 150
wild yeast
   acid washing for removing, 64
   definition of, 63
   genus *Saccharomyces* as, 64
   killer factor produced by, 63
wort
   aeration, 116
      and control of, 118
   boiling, 101
      energy-saving alternatives to, 153
   breaks formation, 48
   Ca^{2+} added to, 70
   kettle, 79
   oxygen addition to, 133
   phosphate addition, 70
   phosphoric acid dissociation, 15
   separation
      and last runnings, 151
      spent grains from, 151

yeast, 114–130
   acid washing of, 116
   addition of oxygen to, 133
   and fermentation, 114
   assessing the quality of, 117, 120
   flocculation, 3, 10
   genetic modification of, 128
   growth of
      measurement of, 116
   in suspension, 116
   nitrogen metabolism in
      fundamentals of, 123
   quantification
      and hemocytometer, 117
   Saccharomyces cerevisiae, 114
   spread-plating of, 65
   taxonomy, 115
      Saccharomyces carlsbergensis, 115
      Saccharomyces cerevisiae, 115

zymocide, 63

Printed in the United States of America